Sociology of Science

Sociology of Science

A Critical Canadian Introduction

Myra J. Hird

OXFORD

UNIVERSITY PRESS

OXFORD
UNIVERSITY PRESS

Oxford University Press is a department of the University of Oxford.
It furthers the University's objective of excellence in research, scholarship,
and education by publishing worldwide. Oxford is a registered trade mark of
Oxford University Press in the UK and in certain other countries.

Published in Canada by
Oxford University Press
8 Sampson Mews, Suite 20,
Don Mills, Ontario M3C 0H5 Canada

Library and Archives Canada Cataloguing in Publication

Hird, Myra J.
Sociology of science : a critical Canadian introduction / Myra J. Hird.

(Themes in Canadian sociology)
Includes bibliographical references and index.
ISBN 978-0-19-542989-3

1. Science--Social aspects--Textbooks. I. Title. II. Series: Themes in Canadian sociology

Q175.5.H57 2011   303.48'3   C2011-902648-1

Cover image: Josh Mitchell/Getty

Oxford University Press is committed to our environment. The pages of this book have
been printed on Forest Stewardship Council® certified paper.

Printed and bound in Canada.

1 2 3 4 — 15 14 13 12

Contents

Boxes and Figures

Acknowledgements

I thank Lorne Tepperman who suggested I write this book. Lorne's support and encouragement throughout this project was invaluable. I am indebted to Peter Chambers whose patience and enduring support throughout enabled me to see this project to completion. I also thank the three anonymous reviewers of an earlier draft of this manuscript. Their comments led to important additions and revisions. I am grateful to the genera Research Group (gRG) for stimulating conversations about sociology, science, and science technology. I am especially grateful to Sandra Robinson and Rebecca Scott for carefully reviewing earlier chapters of this book, and to Sandra for her invaluable assistance in getting the book through the revision process. I also thank my Socy363 Science, Technology, and Society 2009–2010 class for being a critical audience for rough drafts of this book.

I dedicate this book to Eshe, Inis, and Anth,
who are my antidote to this other life.

1

Science, Technology, and the Sociological Imagination

Learning Objectives

In this chapter we learn:
- ⊛ Sociology applies its 'imagination'—seeing the strange in the familiar, and individuality in social context—to science and technology;
- ⊛ Science and technology are a heterogeneous assemblage of local and global practices. Science incorporates and rejects aspects of indigenous, non-western, post-colonial, and feminist knowledge;
- ⊛ Sociology is interested in the ways science and technology are constructed through political, economic, cultural, technological, and methodological practices; and
- ⊛ Sociology critically examines the idea and ideal of modern science.

Science and Technology in Canadian Society

Science defines itself as theoretical, empirical, and practical knowledge about the world derived from the scientific method. Individuals, laboratories, universities, industrial complexes, and institutions across our planet produce scientific knowledge. Science and technology have transformed the way in which we see our own bodies, life and death, relationships with humans and non-humans, our local earthly environment, and the universe. In short, science and technology have profoundly changed our lives. As Steve Shapin writes:

> Few observers disagree when it is said that science has changed much about the way we live now and are likely to live in the future: how we communicate, how long we are likely to live and how well, whether the crucial global problems we now confront—from global warming to our ability to feed ourselves—are likely to be solved, indeed, what it will mean to be human. (2010: 381)

Given the significant influence of science and technology on society, sociology is keenly interested in exploring what science and technologies are; how they operate; and their myriad impacts on people, social structures, and society as a whole. A number of innovative interdisciplinary programs in science and technology studies have developed within Canada including York University's Science and Technology Studies program; the University of Toronto's Institute for the History and Philosophy of Science and Technology;

the University of British Columbia's Department of Philosophy's emphasis on science and technology studies; the University of Alberta's interdisciplinary program in Science, Technology and Society; and Queen's University's genera Research Group. And while not necessarily part of formalized programs or groups, a growing number of sociologists across Canada study aspects of science and technology studies, including political economies of science and technology, philosophy of science, biotechnology, scientific innovation, history of science, and information and communication technology to name only a few. In 2007 the Social Sciences and Humanities Research Council of Canada funded a seven-year project to foster interdisciplinary science and technology studies within the social sciences and humanities. The project, entitled Situating Science: Science in Human Contexts, is organized around the themes of historical epistemology and ontology, material culture and scientific/technical practices, scientific communication and its publics, and geography and the sites of knowing (see the website at the end of this chapter). A number of journal special issues are also devoted to the study of aspects of science and technology studies in Canada, such as the 2006 issue of *Scientia Canadensis: Canadian Journal of the History of Science, Technology and Medicine*, as well as a growing number of books on various related topics, such as John Vardalas's *The Computer Revolution in Canada: Building National Technological Competence* (2001) (see also Powell, 2008; Jones-Imhotep, 2004, 2005, 2009; Jarell and Ball, 1980).

This book provides an overview of how sociology approaches science and to a lesser extent, technology. It concerns how science developed as a set of theories about *what* we know, and *how* we know. How does science define a fact? Can science prove anything? How do science and technology work in practice? Do science and technology shape our values, and if so, in what ways? What is the impact of certain kinds of science—stem cell research for example—on society? What is the impact of culture, politics, economics, and technology on science? These questions, and many more, animate this book.

Science, Technology, and the Sociological Imagination

In 1959 C. Wright Mills published a slim volume entitled *The Sociological Imagination*. In this popular book, Mills sets out what he terms the 'promise' of sociology. Sociology, according to Mills, 'enables its possessor to understand the larger historical scene in terms of its meaning for the inner life and the external career of a variety of individuals' (1959: 5). Mills argues we cannot understand our own experiences in life outside of the society within which we live. He provides three tools that help us better understand phenomena in this distinctive sociological way. He wrote about the ways in which we need to (1) understand *individuality in social context*; (2) see the *general in the particular*; and (3) see the *strange in the familiar*.

To illustrate what Mills means by the sociological imagination, let's briefly consider women's role in science. Looking through the history of science, we immediately see the relative lack of women in science occupations compared with men. This is a familiar situation, but as sociologists we want to view this as a strange phenomenon in need of explanation. In other words, we do not take the absence of women in the sciences for granted. We want to examine the societal structures and forms of power necessary to deter women from advancing in science. Were we to study a particular female scientist from a sociological perspective, we would understand her experiences of practising science within a laboratory, field site, university, and so on, as common to women practising science within a male-dominated environment. Again, we would focus on the ways in which institutions, organizations (such as education), and power operate in order to discourage women from pursuing careers in science. The promise of sociology is to render visible general patterns of social life in the behaviour of individuals, show how society acts differently on different people, and demonstrate how society categorizes groups of people and then assigns different meanings and life opportunities to these groups.

Mills's sociological imagination provides sociologists with a starting point for investigating what science and technology are, how they work, and what their relationships are with past and contemporary societies, as well as their possible futures. In its most general sense, it means that sociologists approach science and technology from a critical perspective. We are interested, for instance, in how power circulates through science and technology, government and industry funding priorities, laboratory practices, political economies of access to science and technology, how science and technology shape our understanding of what it means to be human, and so on.

Sociological analyses of science often differ, sometimes radically so, from received views of science. That is, insofar as different communities of people tend to either wholeheartedly endorse or eschew science and technology—we find polar opposite opinions on many issues regarding science and technology such as stem cell research, cloning, genetically modified foods, and Facebook—sociology is interested in looking at *what* information circulates about a given scientific or technological innovation, *how* it circulates, and what the often complicated and contradictory *implications* of a given scientific or technological innovation are.

Sociology may be witnessing a shift in its overall approach to science and technology from one solely concerned with critique, to analyses that engage with scientific and technological knowledge. Critiques of science, write Mackenzie and Murphie (2008), include critical theory analyses of scientific rationality as well as social constructionist analyses of the processes of scientific knowledge and objects. Chapter 2 focuses on some of the major sociological critiques of science. Extraction is interested in using scientific concepts within philosophy. For instance, Massimiano Bucchi (2004) uses the genetics research terms 'cross-talk' and 'double helix' to explain communication between science specialists and members of the public. Extraction is typically done from science to social science. Finally, engagement attempts dialogue, conversation, and collaboration with science: 'it engages with science-in-the-making and it has had to formulate questions about how to live in or with science collectively' (ibid., 89; see also Barry, Born, and Weszkalnys, 2008). In order to engage with science and technology, sociologists must use the skills we have gained from our own discipline, and also learn enough of the relevant fields of science to understand the basis of scientific claims about phenomena. For this reason, scientists who retrain in the social sciences produce some of the most compelling engagements with science.

This book is intended, as much as possible, to engage with scientific knowledge. It does so from the perspective that science and technology are socially defined by the entanglement of material, economic, political, and cultural relations. Chapter 2 details how science is social relations by exploring, for instance, how science is embedded in thought collectives and paradigms and the acknowledgement that nonhuman entities also engage with science and technology. Chapter 3 explores the social aspects of how scientists actually perform science through educational training, laboratory experiments, grant writing, and the dissemination of scientific knowledge. Chapter 4 further illustrates science and technology as social relations by focusing on quantum theory and molecular biology. Chapter 5 explores science, technology, and power. Political economies of science, feminist science studies, anti-racist science studies, and (dis)ability science studies explore how power operates in specific ways through science and technology. Chapter 6 comes full-circle, addressing how science and society are indelibly tied through values, credibility, and trust.

The Idea and Ideal of Modern Science

Modern science is both an idea and an ideal. The idea of modern science begins with a particular telling of history as a series of events that occurred in a linear progression towards more complete and better knowledge of the universe, and ever better technologies to help people live enhanced lives. Modern science ostensibly began at the end of the Dark Ages, an historical period so-named because of the ascendancy of religions that eschewed scientific inquiry, because of the Black Death that killed a staggering fifty per cent of Europe's population, and because of a number of successive wars (Ede and Cormack, 2004). During the Renaissance, natural philosophers—the precursor to the modern scientist—showed a keen interest in nature and began to use observation, experiments, and deductive reasoning from hypotheses. This set the stage for what is known as the Scientific Revolution—advances in astronomy and physics—that led to the Enlightenment.

The ideal of modern science is that through rational and objective means science accumulates reliable and valid knowledge about the universe (Goldman, 2006). As an ideal, modern science asserts that it is the best way to derive knowledge, and as such, is unique amongst all other forms of knowledge and means of obtaining knowledge.

While neither denying nor minimizing the innovations made by science and technology throughout history, the sociology of science is critical of modern science as a particular way of telling history, and of constructing beliefs about knowledge and truth. Modern science was, as Steve Shapin argues in his book of the same name, *Never Pure* (2010). That is, the history and sociology of science reveals science and technology to be the product of messy, incidental, disordered social forces and events, and not the discovery of a 'deterministic, fully law-governed, and potentially fully intelligible structure that pervades the material universe' (Dupré, 1996: 2). Sociologists of science question the basic tenet that modern science has a privileged claim to truth. The remainder of this section traces the history of modern science as both an idea and an ideal.

The Idea of Modern Science

The development of modern science is narrated through a series of western-defined periods of history: Antiquity, Dark Ages, Renaissance, Enlightenment, Scientific Revolution, and Modernity. It is important to note that non-western cultures have not classified the history of science in this same way. The western idea of modern science typically begins with Antiquity, an historical period during which humans learned how to communicate orally, and later to write, providing the means by which generations passed down information (Ede and Cormack, 2004). Ancient peoples also collected astronomical information, information about human and animal physiology, and alchemy (which later developed into chemistry).

The earliest Greek philosophers, known as the pre-Socratics, were keenly interested in understanding physical phenomena. Thales, often referred to as the father of science, sought non-supernatural explanations for natural phenomena like the weather. His student, Pythagoras, developed mathematics. Leucippus and his student Democritus, developed the theory that matter is composed of tiny units called atoms. Aristarchus theorized that the planets revolve around the sun (known as the heliocentric system). Eratosthenes accurately calculated the circumference of Earth. Hipparchus catalogued the stars.

The famous Antikythera mechanism was an analog computer used to calculate the planet's positions (and which surely provides the idea for the alethiometer in Philip Pullman's *His Dark Materials* trilogy). Hippocrates described a number of diseases and other medical conditions of the body. He also wrote the Hippocratic Oath, which states doctors must practise medicine ethically. Euclid introduced a number of mathematic concepts such as theorem and proof that are still in use today. Archimedes worked out the number pi, amongst other accomplishments. Pliny wrote an encyclopedia of nature, and Theophrastus described animals and plants in the first taxonomy.

In Canada, the idea of science and technology begins with the arrival of the first peoples. Native and Inuit peoples accrued knowledge about their environment through generations of first-hand inductive, deductive, and trial-and-error methods. They developed various technologies focused on obtaining food, making clothing and shelter, and travelling long distances, including the canoe and paddle, the snowshoe, the igloo, and pemmican (a high protein food made from meat and fat that can be stored for long periods of time) (Wright, 1999, 2001). Native peoples also invented many agricultural techniques, which enabled communities to settle in one place rather than live nomadically. Besides developing an extensive and in-depth understanding of the local environments in which they lived, Native and Inuit peoples also invented and developed significant knowledge about medicine and health care (ibid.). Colonial settlers adopted many of these innovations while others were actively suppressed in a systematic attempt to replace native culture with English and French colonial methods of hunting, farming, transportation, and the like (Edmonds, 1973).

Colonial erasure and incorporation of native scientific knowledge and technological innovation is part of modern science's general response to non-western knowledge and traditions. The origin of western science is associated with Greek history: Aristotle, Ptolemy, Galen, and so on. Eastern science is associated with Islamic countries, China, and India. Modern science and western science have become synonymous, affecting both an erasure of the profound (and continued) impact of non-western knowledge on science and technology, and producing a hierarchy of knowledge such that knowledge derived in the west is valued more highly than non-western knowledge.

In what is now known as Iraq, the peoples of Mesopotamia recorded their observations of the physical world with numerical notation. Babylonians noted star, planet, moon, and sun motion through the sky. Indeed, it is from this period—around 3500 BCE—that we get our seven-day week, lunar month, and solar year (Aaboe, 1991). As Asger Aaboe writes, 'all subsequent varieties of scientific astronomy, in the Hellenistic world, in India, in Islam and in the West—if not indeed all subsequent endeavor in the exact sciences—depend upon Babylonian astronomy in decisive and fundamental ways' (1974: 21). Ancient Egypt also developed early forms of medicine that included examination, diagnosis, treatment, and prognosis—what medical students in Canada learn today as SOAP: Symptom, Objective, Assessment, and Plan (Lloyd, 1979).

During antiquity, natural philosophers were more focused at the time than their Greek compatriots on what modern science calls the scientific method:

deriving knowledge through empirical research. Indeed, Ibn al-Haytham—who wrote *Book of Optics* in 1000 BCE—is often credited as the world's first scientist because he systematically used the scientific method to advance our knowledge of optics (Steffens, 2006). The focus on experiment led to developments in mathematics (the term algebra is derived from *al-jabr*, for instance), astronomy, chemistry, and other sciences. According to David Tschanz (2003b), Ibn Sina was the first person to invent clinical drug trials in medicine.

Peoples in India and China made scientific discoveries and generally advanced science in a number of ways, including mathematical measurement (the ruler, for instance), metallurgy (refining metals), astronomy, and

---

BOX 1.1 ⁛ AL-RAZI AND IBN SINA: FOREBEARS OF MODERN MEDICINE

In 'Arab Roots of European Medicine' (2003a), David Tschanz describes the development of medical research and practice in Persia. Following the collapse of the Roman Empire, a number of scholars from Plato's Academy immigrated to Persia and began to work at the university at Jundishahpur. In Europe, Christianity focused on caring for the soul as much as the body, and monks provided a supportive surrounding within which sick people would either get well or die, both according to God's plan. In Persia, hospitals emerged in conjunction with medical schools, libraries, and medical student training.

Al-Razi (known as Rhazes in the west) became a doctor at the age of 40. He was concerned with many branches of medicine, including fevers, kidney stones, scabies, smallpox, measles, toxicology, and the climate's effect on health. He wrote a staggering 237 books, including *The Diseases of Children*, which some argue makes Al-Razi the first pediatrician. In his most well-known book, *The Comprehensive Work*, Al-Razi urged doctors to practise medicine based on observation rather than received authority. He also wrote about the need for high professional standards, and the importance of trust within the doctor–patient relationship.

Ibn Sina studied medicine in Persia some time after Al-Razi's death. Ibn Sina was only 16 years old when he began studying medicine, and wrote a number of key medical texts, including *The Book of Healing* and *The Canon of Medicine*. The latter opus amounted to a comprehensive exposition of all medical knowledge known to date, and included information on anatomy, physiology, diet, the effect of climate on health, ulcers, breast cancer, epilepsy, diphtheria, leprosy, rabies, diabetes, and gout, as well as how to diagnose and treat patients. Ibn Sina developed successful surgery for cataracts and hernias, filled teeth with gold, and invented drug trials. His work was considered the gold standard of medical diagnosis and treatment and was influential well into the nineteenth century.

trigonometry. China is well known for its Four Great Inventions: the compass, gunpowder, paper, and printing (Needham, Robinson and Huang, 2004).

The fall of the Roman Empire during the late 400s CE marks the end of the classical antiquity period and the beginning of the Middle Ages (roughly the fifth through to the sixteenth centuries). This is generally viewed as a period in which knowledge about the natural world—medicine, astronomy, mathematics, and so on—was lost as libraries, universities, and other institutions of learning were destroyed during successive invading conquests throughout Europe and the Middle East. This is not to say that natural philosophers did not make new discoveries or innovations during this long period; rather these developments took place within the classical Greek tradition of deduction based on first principles, rather than the scientific method. Massive events like the Black Death effectively halted scientific inquiry.

The resumption of scientific exploration during the Renaissance (beginning approximately in the twelfth century) saw a gradual shift away from classical Aristotelian science, which was based on deductive reasoning. During this period the Church and Monarchy ruled European countries. The vast majority of Europe was illiterate and education in the modern sense was prohibited. Democracy did not exist: the sovereign (King or Queen) held the power of life and death over its people. The Church supported a literal interpretation of the Bible, including heaven and hell as physical places. The Great Chain of Being conceived of an ordered and structured God-given universe in which God commanded from the top, followed by angels, humans, animals, plants, minerals, and the rest of the material world (dirt and so on). The order of the universe, according to this theory, was hierarchical with each element (humans, animals, and so on) occupying a definite, fixed, and purposeful place within the structure. Christian knowledge was founded upon revelation and tradition. God revealed to the sovereign His divine plan, and clerics (priests or pastors) ministered to the people according to this grand plan. There was no scope for discussion or negotiation between the people and the Church, since God Himself decreed this order.

Things began to dramatically change in the fifteenth century when Copernicus introduced a heliocentric cosmology. Up until that point, both the Church and Greek astronomers—most famously Ptolemy—maintained that the sun and planets revolved around Earth. Copernicus figuratively decentred Earth, replacing it with a sun-centred universe. The German mathematician and astronomer Johannes Kepler improved upon the refracting telescope developed by Copernicus and further refined Copernicus's cosmology. Galileo Galilei made significant advances in telescope technology, with which he discovered sunspots and the moons of both Jupiter and Saturn.

Galileo's confirmation of Copernicus's sun-centred universe brought him into direct confrontation with the Catholic Church and, as a result, he spent a significant proportion of his life under Church-ordered house arrest. Galileo's

BOX 1.2 ❀ GALILEO AND THE CATHOLIC CHURCH

The Catholic Church took strong objection to Galileo's endorsement of the Copernican sun-centred account of the solar system. It was Galileo's insistence that the sun-centred account was the truth, that it was certain, universal, and necessary, rather than confining the theory to a system that predicted the movement of the planets with greater precision than Ptolemy's system, that got him into trouble (Goldman, 2006).

The Bible states that the sun normally moves and that it stood still for Joshua in Aijalon. Copernicus's system requires Earth to move and the sun to remain still. Galileo was essentially saying that the Bible should be read as a figurative document (and thus open to interpretation) rather than literally. In the context of the Church fighting the Thirty Years' War (from 1618–1648) in which millions of people were killed in the name of religion (Protestantism versus Catholicism), the Church understandably took exception to Galileo's ideas, and he was placed for years under house arrest.

It is no accident that the motto of the (British) Royal Society, founded some 300 years ago, is *nullius in verba*—take nothing on authority.

technical breakthroughs (such as the telescope) combined with his approach to studying the physical world through meticulous and detailed observation, led to a dramatically different conception of the world, and humanity's relationship with it. By displacing Earth (i.e., humans) from the centre of the universe, Galileo was effectively displacing God. This meant divine purpose was peripheral in the cosmology of the scientific revolution: the physical world obeyed natural laws, not those created by God.

The Renaissance took place within the context of massive political upheaval in Europe initiated by the Protestant Reformation, the fall of Constantinople, and the exploration and colonization of the Americas by Europeans. These political and religious shifts, combined with complex economic and cultural developments, led to the Scientific Revolution and the Age of Enlightenment.

During the Renaissance, colonial settlers, primarily from Britain, Scotland, and France, ventured to Canada where they envisioned a land of economic and political opportunity. Settlers saw opportunities for extensive land development, agriculture, hunting and fur trading, fishing, trade, and so on. In order to extend their purview over this vast region, colonialists concentrated on mapping the landscape, and figuring out ways to utilize and control the environment (including Native peoples).

Colonialists brought with them various technologies, including the wheel, shipbuilding, sailing, as well as water (mills) and animal power (in the form of oxen and horses) to further develop agriculture (Levere, 1974). They also introduced different farming, fishing, and hunting techniques and technologies, eventually forcing Native peoples and Inuit to abandon their own tried and tested agricultural, hunting, and fishing practices. Colonial explorers such as Jacques Cartier and Samuel de Champlain are famous for charting Canada's coasts.

It is interesting to note the word 'scientist' did not appear in written accounts until well into the nineteenth century, when William Whewell coined it in his book *History of the Inductive Sciences* (1837). Before that time, Whewell and others referred to what we now call scientists as natural philosophers.

People's curiosity about nature was facilitated during the Enlightenment by a number of factors. Merchants and explorers were charting lands and oceans far away from 'old world' homelands (Britain, France, Belgium, Spain, and so on). Explorers and naturalists brought back curiosities—the Lord's prayer written on a grain of rice, a two-headed snake, elephant tusks, and kangaroo tails, for example—and wealthy people purchased and collected these curiosities to put on display in what was known as a Cabinet of Curiosities, Cabinet of Wonder, or Kuntskammer (in German) (Daston and Park, 1998). These so-called cabinets were actually rooms; they were encyclopedias of sorts because they focused on collections of diverse objects. As much as cabinets of curiosities represented power (they were so expensive to maintain and expand that only the wealthiest people could afford them) and wonder (demonstrating the never-before-seen diversity of objects in the world), they also represented a key feature of the Enlightenment ideal, which was the demonstration of control over the environment. Cabinets of curiosities represented a sort of microcosm of a modern world that orders, classifies, and thereby controls objects.

The development of mechanized printing also contributed to people's curiosity about the natural world and science. Before the Enlightenment, few people could read. Access to the Bible, which was typically written in Latin, was through priests who read passages out loud to congregations. As Sandra Robinson notes:

> [mechanized] printing had a crucial role in expanding scientific knowledge. . . . [It] was critical to the circulation of scientific atlases and treatises that expanded the reach of science beyond academic élites solidifying early modern scientific enterprise in the eighteenth century. . . . [T]his advancement in capability and capacity through mechanized printing made scientific atlases more accessible and increased their circulation. Printing effectively publicized science. . . . Thus, by the time the notion of objectivity became critical to science, printing and distribution had become critical to scientists. (2010: 115)

Jesuits who emigrated from France founded the first college, the College de Quebec, in 1635. By the late 1800s the G-13 or group of 13 large Canadian universities had been established. These universities taught courses in natural history, which later became zoology, botany, biology, and geology; and natural philosophy, which evolved into physics. The Association of Universities and Colleges of Canada was formed in 1911 (Harris, 1976). A number of societies, such as the Botanical Society of Canada, were founded in the 1800s as well, almost always mirrored on their earlier established counterparts in the old worlds such as England and France. In the late 1800s and early 1900s, a number of government scientific research organizations were founded, such as the Royal Society of Canada (1883), the Geological Survey of Canada (1841), and the Biological Board (1912) (Berger, 1996). The National Research Council of Canada (NRC) was established in 1916 (Eggleston, 1978). The Canadian Medical Association was founded in 1841, helping to establish medicine as a more formalized institution. It was around this time that Charles Fenerty invented newsprint made from wood pulp, and Abraham Gesner invented kerosene (Dewalt, 1995). Mathew Evans and Henry Woodward invented the incandescent electric light, and sold the patent to Thomas Edison (Mayer, 1997). Canadian engineer Sandford Fleming's proposal for a universal time system was accepted at the International Meridian Conference in 1884 (Thomson, 1978).

Research laboratories were introduced into Canadian universities in the early 1900s (Harris, 1976). The physics laboratory at McGill University was the site of Ernest Rutherford's discovery of the atomic nucleus, for which he won the Nobel Prize in 1908. The Connaught Laboratories at the University of Toronto was the site where Frederick Banting and Charles Best discovered insulin, for which they won the Nobel Prize in 1923. Alan Brown, Fred Tisdall, and Theo Drake invented Pablum (pre-cooked baby cereal) at The Nutritional Research Laboratory at Toronto's Hospital for Sick Children in 1930 (MacDermot, 1967). This particular hospital laboratory also produced research showing the health benefits to children of adding vitamin D to milk, and constructed over 30 iron lungs for children suffering from polio. It was also around this time that Robert Mawhinney invented the dump truck, and Joseph Bombardier invented the snowmobile, and later founded the Bombardier Corporation. Canadian scientists and inventors were also involved in the invention of a number of weapons, including the Ross rifle and gas mask, and contributed to the development of high velocity artillery, radar, the proximity fuse, and submarine detection equipment, amongst other weapons (Morton, 1990).

The Second World War provided the impetus for significant changes in science and technology research in Canada. Some areas of science witnessed exponential growth in government and industry financial support, such as physics (initiated by the profound political and cultural significance of the detonation of the two atomic bombs on Japan). The Canadian government devoted significant funds to developing atomic energy (at the

Chalk River Laboratories in eastern Ontario, amongst other places) (Bothwell, 1988). Other disciplines experienced significant decreases in funding (botany, for instance) as the government focused its monetary resources on research and technology development it determined would most directly benefit the war effort.

The Second World War also ushered in more pervasive, complex government- and industry-shared research funding and structures. Scientists, funding agencies, and provincial and federal governments began to create Big Science—research that involves multiple (often international) laboratories, researcher teams, industry partners, multi-million dollar grants, and strong economic directives. Big Science produced atomic research reactors, space satellites, and radio telescopes, as well as transnational medical research projects (ibid.). Big Science has an unclear relationship with what is known as pure science, or science research directed towards increasing our understanding of how things work rather than any particular application. The direction of funding suggests a shift away from pure science towards more applied science, seen as delivering greater promises of short- and medium-term economic gain.

Post-Second World War, universities across Canada showed significant increases in student enrolment as baby boomers entered young adulthood. Campuses expanded, new programs, departments, centres, and institutes were created. Disciplines such as biology, chemistry, astronomy, geology, and physics expanded, financially backed by provincial and federal governments, universities, and industry. It was during the 1970s that Keith Downey and Baldur Stefansson developed canola oil, which is very widely used in western diets today. It was also during this boom time for science and technology in Canada that the Connaught Laboratories in Toronto, in partnership with a US laboratory, created the world's first polio vaccine and what became known as the Toronto Method, which enabled large-volume production and mass vaccination of the Canadian population (MacDermot, 1967). Influenza, measles, and freeze-dried smallpox vaccines were later developed. During this time, more associations, government agencies, and corporations were established. In 1952, for instance, The Ontario Heart and Stroke Foundation, was established. The 1950s also witnessed the development, in Toronto, of the artificial pacemaker for heart patients. Working in the United States, Canadians also invented the alkaline and lithium batteries, and the charge-coupled device.

The growth of science in Canada largely came to halt in the mid-1980s with major government cutbacks to universities. Since that time, universities have experienced a slow recovery of science funding. This said, Big Science and the focus on economic gains through science and technology continued apace. In 1989, for instance, the three largest federal funding agencies—the Medical Research Council of Canada (later re-named the Canadian Institutes of Health Research or CIHR), the Social Sciences and Humanities Research

Council of Canada (SSHRC), and the Natural Sciences and Engineering Research Council of Canada (NSERC)—established as the Networks of Centres of Excellence (NCE) with the explicit goal of facilitating the commercialization of science and technology research in Canada.

Several Big Science projects are notable during this period. The Sudbury Neutrino Observatory (SNO) studied neutrinos from 1999 to 2006. The Canadian Light Source Synchrotron at the University of Saskatchewan at Saskatoon began in 2004. One notable fact about this research site is that it is operated by CLS Inc., which is a not-for-profit corporation. Similarly unusual is the Perimeter Institute for the study of quantum mechanics and relativity in Waterloo, which was created by a single individual (the person whose corporation—Research in Motion—created the BlackBerry). The Canadarm, or Remote Manipulator System devices, research and technology produced in Canada in co-operation with the United States for the NASA space program began during this period as well. In 1999, the world's first cloned goats were produced at McGill University, and in 2001 researchers at the University of Montreal cloned three calves. Path-breaking research into stem cells has recently taken place at the University of Toronto. Researchers at the National Microbiology Laboratory (NML) in Winnipeg, Manitoba first cracked the genetic sequence of the H1N1 virus. And the Robarts Research Institute at the University of Western Ontario's recommendation of a daily aspirin dose to decrease the risk of stroke has become a standard of care. In 2008, the ZENN (Zero Emissions, No Noise) electric car, built in Saint-Jerome, Quebec, was released to the international market.

Currently in Canada, the bulk of science research takes place within university settings (40 per cent) as opposed to government laboratories (ten per cent). In 2007, 10.4 billion dollars was invested in university research, and this research is estimated to have contributed 60 billion dollars to the Canadian economy. Canada ranks sixth worldwide in terms of scientific articles published and cited. The 2009–2010 federal budget allocated over 10 billion dollars to scientific research and development, at the same time that the government made budgetary cuts to scientific research granting agencies such as the Canadian Institutes of Health Research, the Natural Sciences and Engineering Research Council of Canada, the Social Sciences and Humanities Research Council of Canada, and the National Research Council of Canada.

The Ideal of Modern Science

Modern science is not only an idea; it is an ideal. The ideal is that science provides the best approach to achieve knowledge that is certain, universal, necessary, and true (Goldman, 2006). A key aspect of this ideal is that science is a unified system. Sandra Harding outlines four assumptions necessary to this unity-of-science claim:

. . . there exists just one world, one and only one possible true account of it, and one unique science that can piece together the one account that will accurately reflect the truth about that one world . . . [and] that there is a distinctive universal human 'class'—some distinctive group of humans—who should be taken as exemplars of the uniquely or admirably human to whom the truth about the world could become evident. (1998: 166–67)

During the Renaissance, and particularly during the Enlightenment period, a diverse and growing number of natural philosophers, politicians, and social reformers sought to revise the way western cultures understood the world: from a God-given, divinely controlled, static, and hierarchical system to a world in which humans control their own destinies through their unique capacity to reason.

As we learned in the previous section, Copernicus and his successor Galileo challenged the Christian doctrine of an Earth-centred universe, thus metaphorically decentring humanity from the centre of creation. But to really understand the ideal of modern science, we must turn to a central figure in western philosophy.

Immanuel Kant (1724–1804) was a German philosopher whose writings on human reason profoundly shaped western knowledge and arguably epitomized what the Enlightenment was all about. While Kant is most popularly known for dense works such as *The Critique of Pure Reason* (1787), as a younger man he wrote the delightfully titled book *History and Physiography of the Most Remarkable Case of the Earthquake which towards the End of the Year 1755 Shook a Great Part of the Earth* (1756/1994). The earthquake Kant refers to took place on 1 November 1755. Lisbon, the capital of Portugal, was literally torn apart by three successive tremors that opened up the ground in the city centre, toppled buildings, started fires that raged for days, led to several giant tsunamis that flowed through the city centre, and caused the Targus River to spill its banks and further flood the city. The earthquake was felt in Switzerland, France, and northern Italy; waves from the successive tsunamis reached Holland and Belgium, and across the Atlantic ocean to Antigua, Martinique, and Barbados. All told, the earthquake and its after-effects killed thousands of people and wrecked havoc on a city that took years to rebuild.

As Nigel Clark's (2011) analysis of the impact of the Lisbon earthquake on Kant's philosophy points out, the theodicy of the time required that every natural event be a manifestation of God's master plan. And to be sure, suspected heretics were rounded up and punished. But Kant and others suspected that Lisbon's disaster was not God's plan; that it was rather the work of nature. It's helpful to keep in mind here that natural philosophers studying geology at this time were uncovering compelling evidence that Earth was not formed in a matter of days, as the Church maintained, but rather over eons. Copernicus and Galileo had already suggested a solar system much older than Church doctrine.

Faced with a universe whose every minute machination was not under God's care, the young philosopher determined what has become known as Kant's Settlement. He argued that humans are entirely separate from the universe and, further, that humans' unique capacity to reason allows us to recognize regularity and order in nature. This philosophical move separates humans from the universe, and demands the universe conform to our analyses of it. As Quentin Meillassoux writes, 'instead of knowledge conforming to the object, the Critical revolution makes the object conform to our knowledge' (2008: 117–18). Kant's philosophy not only separates humans from the universe, but also establishes humanity's superiority over nature. While humanity does not itself govern nature, humans are capable (through reason) of creating their own governance.

A central Enlightenment objective was to question tradition and revelation as the bases of knowing, replacing these with principles of deduction, logic, and induction from observation and experiment. It introduced the scientific method (see Chapter 3) as the foundation of knowing. According to the emerging ideal of science, knowledge derived from science would reflect the value of scientific discovery directed towards the benefit of humanity as a whole. The Enlightenment promise was to use knowledge derived from science to free humanity from oppression and to create a superior society. It is for this reason that in 1784, Kant encouraged humanity to 'dare to know'. Sociology's founding theorist, Auguste Comte (1853/2009), devised his Law of Three Stages of knowledge: a theological stage in which knowledge is derived with reference to the supernatural (God); a metaphysical stage in which knowledge is based upon appeals to essences (fundamental attributes), whether human reason or natural laws; and the positive stage, in which causation is replaced by the explanation of facts (see Chapter 2). Comte argued western society had advanced through these three stages, and the Enlightenment enabled the third and final stage.

To give one example of this significant shift, prior to the Enlightenment priests routinely admonished parishioners for being curious about the world because the Church taught that God had made the world and that it was meant to persist without change, for all time. Lorraine Daston (2007) provides an example of French villagers in the 1600s who were curious to see a set of human conjoined twins. Their priest reprimanded the parishioners for being curious about what the Church understood to be an abomination of God's work (i.e., that conjoined twins were the work of the Devil). In order to effectively get around this problem, Francis Bacon—a famous English philosopher, statesperson, scientist, and lawyer who served as both Lord Chancellor and Attorney General of England—distinguished between two kinds of curiosity: curiosity about God remained taboo, but curiosity about the workings of the world represented a natural outcome of God's endowing humans with the capacity to explore and reason. This scientific curiosity about the

---

BOX 1.3 ❄ **DARWIN AND THE IDEAL OF MODERN SCIENCE**

Because the Christian Church maintained the principle that the world had not changed since God created it, the Church was unable to address the widely observed phenomenon of species change. Through over 30 years of painstaking study based on observation, collection, dissemination, and discussions with other natural philosophers, Charles Darwin theorized that species change significantly over time in response to their changing environments. Animal breeding, for instance, selects certain characteristics (for example, in dogs this means long ears, short fur, and the like) that add up over time. Darwin posited that, just as humans select domesticated animal characteristics, nature does this relentlessly with all species. The mechanism, Darwin theorized, is Natural Selection. 'The preservation of favourable variations and the rejection of injurious variations, I call Natural Selection,' wrote Darwin in his monumental opus *On the Origin of Species* (1859). Natural selection works by preserving and thus accumulating very small, inherited beneficial variations in a species.

Darwin's theory of evolution shook the academy and captured the public imagination. In many ways it epitomized the ideal of modern science because it was based on the scientific method of reasoning coupled with testable hypotheses and reliable observation.

---

world led western cultures to further challenge Christianity's account of the world, and add fodder to the ideal of modern science as the conduit of truth.

The Enlightenment witnessed a dramatic shift in the relationship between religion and sovereign forms of governing, and ultimately the dominion of Church over state. In France, the revolution during this time eradicated the King and aristocracy. Significant changes occurred in how people were governed, and social and economic class structures shifted. As a consequence of this period, people changed the ways in which they looked upon the possibilities, scope, and function of scientific inquiry itself.

Sociological Critiques of Science and Technology

Thus far we have learned that modern science, as both idea and ideal, played a crucial role in establishing what historians, philosophers, economists, politicians, and indeed the public, call modern society. Science developed as an idea about how we might come to understand, and better live within, our environment. The development of the scientific method seemed to establish a way to gain valid and reliable knowledge. In theory, all of humanity had access to the powers of reasoning and observation; the Church would no longer circumscribe to the revelation what could be known about the world. Science developed as an ideal about *what* knowledge is valid, and how knowledge *should* be accrued. In short, the Enlightenment promised a better world for all of humanity through reason, with science as its strongest ally and handmaid.

Sociologists are skeptical of these Enlightenment claims. Sociologists who study science and technology argue western society has never, in fact, been modern; that science never replaced religion; and that science has never been pure in the double sense that science has never produced pure knowledge, nor have science and technology purified themselves from power relations.

We Have Never Been Modern

According to sociologist and philosopher Bruno Latour (1993), modernity has largely been an attempt to purify the world by separating humans from the rest of the universe. On one side of this chasm, as modernity would have it, are humans imbued with the freedom to act upon the world. On the other side of this chasm, according to modernity's construction, lies the rest of the universe (animals, rocks, raindrops, and all), which is bound by brute physical laws. As such, the rest of the universe is largely inert, as if waiting for humans to act on it. Or, as Graham Harman puts it, 'this radical split between the mechanical lethargy of objects and the floating transcendent dignity of human subjects is the typical feature of the West during the modern period' (2009: 76).

One of the profound consequences of the Copernican revolution and Kant's subsequent settlement was, as we learned earlier in this chapter, to attempt to enforce a strict and enduring divide between humans and everything else. This has certainly affected the way in which we study phenomena. Whereas the natural philosopher of the twelfth century might, with equal interest and credibility, study fossils and Greek literature, the scientific revolution led to the isolation of humans from all other beings and objects (Goldman, 2006: 231). Academia has certainly felt the effects of this divide, developing into three separate domains: natural sciences, social sciences, and humanities. Contemporary universities exemplify this disciplinary separation in their very geography. As sociology students, your department is likely to be housed in a building with other social sciences—anthropology and political studies, for instance. You are unlikely to bump into molecular biologists in the hallway because they are working in buildings across campus, beside other natural science disciplines.

By and large, sociology has supported this carving up of domains. Adelard of Bath argued natural phenomena might only be explained by other natural phenomena (Goldman, 2006). Emile Durkheim, a founding sociologist, provided a later addendum that social facts must be explained by other social facts (1982). Thus, social scientists have been more interested in 'explor[ing] the processes whereby certain problems come to acquire "real" status at particular moments and in particular contexts' than in determining how real an object is and what sociologists might do about it (Irwin, 2001: 180).

But, as Latour argues, we have never actually been able to purify the world, meaning—as the title of one of Latour's books suggests—*We Have Never Been Modern* (1993). Latour means that the very dualism of nature and culture does not exist. A number of sociologists are joining philosophers and scientists alike in collapsing the divide between objects and humans. Vincent Mosco (2010), for instance, argues science and culture and the arts were never distinct, and indeed have always depended upon each other. Chapter 2 reviews some of the more prominent attempts to challenge modernity's legacy.

Science Never Replaced Religion

As Steve Shapin observes, it is so commonplace as to hardly bear scrutiny to say that 'science made the modern world' (2010: 377). 'It's science,' writes Shapin, 'that's understood to be the motive force . . . of the leading edges of change. It's science that drives the economy and, more pervasively, it's science that shapes our culture' (ibid.). Contemporary North American, Western European, and Antipodean countries largely go about their business as secular societies: religion is reserved to particular spheres of life such as the family and, for the most part, is kept separate from the advancement of society through science. It is also commonplace to regard evolution as largely accepted within the scientific community and academia generally. Even the Catholic Church recognizes evolution as, according to Pope John Paul II, 'more than a hypothesis' and 'an effectively proven fact' (Linder, 2004: 82). (Far earlier, in the nineteenth century, Pope Leo XIII accepted that humans evolved from animals but believed God created the human soul). John Paul, indeed, acknowledged:

> Today, almost half a century after publication of the encyclical, new knowledge has led to the recognition of the theory of evolution as more than a hypothesis. It is indeed remarkable that this theory has been progressively accepted by researchers, following a series of discoveries in various fields of knowledge. The convergence, neither sought nor fabricated, of the results of work that was conducted independently is in itself a significant argument in favor of the theory. (ibid.)

Except, as Shapin points out, there is a lot of evidence to indicate that we do not now, nor have we ever, lived in a scientific world. According to various polls, most Americans—about 90 per cent, in fact—believe in a deity. Slightly less believe heaven is a physical place. And if, as Shapin's synthesis of self-reported polls shows, we only take into account educated people's views, we

BOX 1.4 ❋ IS INTELLIGENT DESIGN SCIENTIFIC?

In the United States, and to a lesser extent in Canada and Europe, the Intelligent Design (ID) versus evolution debate has garnered sustained attention as the latest reiteration of the debate about the separation of religion and state in western societies. Interestingly, ID advocates attempt to invoke science—scientific concepts, scientific language, and references to scientific studies—to bolster their claims. In an important sense, this choice to use science is an implicit acknowledgment of the power that science holds in contemporary western societies. Science, in other words, persuades people.

ID advocates argue evolution is no more true or factual than creationism, and that evolution itself proves there has to have been an overarching creator of nature's complexities. Creationists point out that scientists refer to evolution as a theory. But scientists refer to theories when they are describing a particular structure of ideas. Scientists do not use the term theory in the way the public does: to describe a conjecture or set of assumptions. The National Academy of Sciences defines scientific theory as 'a well-substantiated explanation of some aspect of the natural world that can incorporate facts, laws, inferences, and tested hypotheses' (1999).

ID, Steve Goldman argues, 'violates one of the most fundamental principles of sciences . . . by bringing a supernatural agent into scientific explanation' (2006: 183). In science, only natural phenomena can explain natural phenomena. It's the who-designed-the-designer problem. Moreover, for a hypothesis to be scientific, it must have predictive success. That is, a hypothesis must be able to make predictions that then occur (or at least point to other hypotheses whose predictions will occur). ID offers no research program because it does not have access to the supposed designer's design. Without access to the design, the researcher would have to act as if there were no design at all. Finally, ID arguments are a good example of the importance of engaging with scientific knowledge. In *Darwin's Black Box* (1996), Behe points to a number of complex phenomena in nature—the eye, bacteria flagella—that require the interaction of systems that could not supposedly have evolved separately by random mutation. Proponents of ID argue there must be some force (God) behind these complex phenomena: there must have been a designer. The problem with this argument is that scientists *have* demonstrated how simple systems can self-organize into complex systems. (Check out the website at the end of this chapter).

still find about 84 per cent of university-educated people believe in things like miracles, while less than 17 per cent agree with evolution. And a significant number of scientists themselves believe in a deity. Most members of the public know very little about scientific knowledge itself or the scientific method, let alone how science works in practice. And even scientists are likely to only know a significant amount about their own sub-fields and area(s) of research, rather than the sprawling opus of science per se.

This is not to say that westerners do not also put a lot of faith in science, to cure cancer and AIDS, to solve global climate change and food shortages, and to battle obesity. But even here there is tension between knowledge and moral authority. As Shapin observes:

> Science is our most powerful form of knowledge; it's scientists . . . that are turned to when we want an account of how matters stand in the natural world. But. . .it is not scientists who decide what ought to be done. For those decisions—and there are an increasing number of them that are potentially world-changing—it's politics as usual. (2010: 389)

So, western societies may no longer strictly consider religion as their ethical compass, but they have not turned to science and technology as the indefatigable sages of morality.

Science Has Never Been Pure

Sociologists of science and technology challenge the ideal that science is our best tool for understanding the world. In the first instance, sociologists point out that modern science never purified itself from other sources and modes of knowledge. That is, modern science has always incorporated knowledge and methods and modes of learning about the world, from non-western sources. Yet, as Harding argues:

> No matter how much modern sciences might have incorporated elements of other cultures' concepts and theories about nature, their mathematical and empirical techniques, and even whole bodies of their accumulated navigational, medical, pharmacological, climatological, agricultural, manufacturing, or other effective knowledge enabling prediction, and control of nature, these bodies of knowledge are not counted as 'real science' until incorporated into European knowledge systems. (1998: 166)

The key contributions made by indigenous and non-western peoples in mathematics, astronomy, medicine, chemistry, agriculture, zoology, and other subjects outlined in the previous section provide a flavour of the deeply intertwined histories of western and non-western science-in-the-making. Sociologists emphasize the key point that western science's effacement of these significant contributions, and indeed the very shaping of what western cultures understand to be science, is part of the reason modern science has never been pure.

A second reason that modern science has never been pure is that science is inseparable from power in society. By this, sociologists mean that while rhetorical claims are made in science that at least some research is undertaken purely to advance human knowledge about the world, the vast bulk of science and technology research takes place within the context of neo-liberal consumerist societies and their interest groups. As Shapin wryly summarizes:

> The technologist supplies what society wants; the scientist used to give society what it didn't know it wanted. That's a simplification, but, I think, a useful one: corporations, governments, and the military enlist experts in the natural world overwhelmingly on the condition that they can assist them in achieving useful goals—wealth and power. (2010: 390)

Ursula Franklin—Canadian metallurgist, physicist, pacifist, and winner of Canada's Governor General's Award and Pearson Medal of Peace—wrote *The Real World of Technology* (1992) in which she argues the uses of technology are not preordained but rather result from established structures of power and control. Franklin uses the example of the mechanical sewing machine. Introduced in 1851, this technology was touted with the promise of liberating women from domestic chores. But sewing machines are mainly used in factory sweatshops producing cheap clothing and further exploiting female workers around the globe. Indeed, Franklin argues, technologies may be defined as anti-people, insofar as technologies are viewed as providing solutions, while people are more often seen as a source of problems when labour issues (such as how many hours at a time people can operate machinery safely) get in the way.

Moreover, the history of science and technology briefly outlined in this chapter demonstrates the increasing dependence of science and technology on government and big business funding. University laboratories, where the bulk of scientific research takes place in Canada, require significant ongoing funding which comes mainly from partnerships between government and industry. A consequence is that the benefits of scientific knowledge and technology are unevenly distributed across the globe, tending to favour already relatively enfranchised people (white, European, and North American men) and impede already disenfranchised people.

Governments and industry are, of course, not neutral: both are guided by multiple agendas that include wealth and power (as Shapin points out), as well as values about what is in society's best interest (Aronowitz, 1988). The well-known example of the development of atomic weaponry illustrates how science and technology have never been purified from societal values, politics, wealth, and power.

Up until the First World War, Classical Studies—languages such as Latin, as well as art, literature, history, and philosophy—were always considered to be the penultimate fields of scholarly study. After the war, science—and particularly physics—began its ascent in popularity. From the 1920s to 1940s, physics dominated. Whereas before the war scientists tended to work either in isolation or in small groups, the American Chemical Warfare Service taught scientists the value of big research groups and big funding for the advancement of research (Ede and Cormack, 2004). Whereas physicists before the war could work amongst other physicists at a few places—the Cavendish Laboratory or the laboratory at Cornell University, for example—after the war, big private funding (Rockefeller Foundation, Carnegie Foundation, and so on) combined with vigorous new interest in universities to study physics. Theoretical and experimental physics—led by Albert Einstein, Lise Meitner, Max Planck, Werner Heisenberg, Erwin Schrödinger, and Otto Hahn amongst others—took hold of the scientists', and later the public's imagination.

During the 1930s and 1940s the fear that Germany was attempting to devise some kind of super weapon made physicists keenly interested in better understanding radioactivity. When scientists learned that Hitler's Germany had invaded the Belgian Congo—the most well-known source of uranium—they felt Germany's potential to create an atomic bomb was both real and imminent. This threat led Albert Einstein to write to American President Franklin D. Roosevelt in 1941, an act which ultimately initiated the Manhattan Project. The Manhattan Project was a secret research project whose aim was to build a nuclear bomb before the Germans. Science and politics changed the course of history.

When the first bomb exploded in what was called the Trinity Test in July 1945, it represented a simultaneously remarkable advancement in new scientific knowledge about the world and a profound ethical dilemma. By this time Germany had been defeated and the last remaining enemy, Japan, was losing. But Harry S. Truman, Roosevelt's presidential successor, wanted to test the bomb in order to provide a categorical response to Pearl Harbor, avoid further American casualties in the Pacific, and prevent the USSR from gaining any section of Japan in the case that it was partitioned after the war, as Germany had been after the First World War. On 6 August and 9 August, two bombs were dropped by America on Hiroshima and Nagasaki, killing more than 200,000 people, most of whom were civilians. These devastating bombs led swiftly to an arms race between the USA, Britain, and the USSR, all developing further nuclear weapons such as the hydrogen bomb.

Physicists along the route to building these weapons of mass destruction several times petitioned governments to cease the arms race, and tried to convince the public of the folly of pursuing this type of weapon development through techniques such as putting a doomsday clock on the cover of their *Bulletin of Atomic Scientists*, which was set in 1947 at seven minutes to midnight (midnight equaling the end of human life), and in 1953 after the USA and USSR tested fusion weapons (about 700 times the power of the atomic

---

BOX 1.5 ❋ HIGHWAY OF THE ATOM

Peter van Wyck (2002, 2005, 2010) provides a superb analysis of the physical, social, political, and cultural impact of uranium mining in northern Canada by Native peoples. Van Wyck traces the 'highway of the atom' starting with the uranium mined by the Dene of Great Bear Lake in Canada's Northwest Territories. The uranium travelled through northern Alberta and was transported to a refinery in Port Hope, Ontario. It was then taken to the United States and the Manhattan Project, where it was combined with ore from the Belgian Congo and Colorado, and made into the bombs dropped on Hiroshima and Nagasaki.

The Dene experienced many losses of life due to radiation-related sickness (the mine site was named Port Radium) that continue to today. And in an extraordinary turn, the Dene concluded that they had—unknowingly—participated in the genocide of Japanese people. This led the Dene to a remarkable journey to Japan, where they officially apologized to the Japanese people and commemorated the staggering loss of life due to the atomic bomb. The Dene's extraordinary approach invites questions about individual, community, and societal relations of responsibility with science and technology. See Peter van Wyck's *The Highway of the Atom* (2010).

---

bomb dropped on Hiroshima, and capable of devastating entire countries), the clock was set at two minutes to midnight.

The Atomic Age brought science and society into sharp relief. What role should scientists play in determining national security and military policy? For some, scientists are duty-bound to work in co-operation with the military when national security is threatened. For others, scientists must voice concerns against military weapons development because only scientists can appreciate the consequences of these developments. Whatever the view, as Harding (1998) points out, disassociating modern science from its material, political, economic, cultural, and social effects is increasingly difficult to defend. She argues the ideal of modern science has had the effect of internalizing the benefits of scientific and technological change, and externalizing their costs as misapplications rather than the scientific processes themselves. Sociologists, by contrast, argue science, technology, and values cannot be dissociated in this way, and this is the subject of the final chapter of this book.

Sociologists argue modern science was never pure in a third sense. Sociological analyses reveal the processes of scientific and technological research to be inherently and necessarily messy, incidental, and disordered rather than orderly and logical, as the ideal of modern science maintains. This book is devoted to detailing sociological analyses of science and technology as social enterprises.

Summary

This first chapter provides a brief synopsis of the history of science and technology in Canadian society. Long before French and British explorers tracked through Canada, Native and Inuit peoples had developed ways of finding out about and understanding nature, as well as sophisticated and practical technologies to enhance their survival in a variety of climates and environmental conditions. Explorers brought with them various European techniques for charting and evaluating the geography, different agricultural techniques, and a Christian and western-based worldview of the cosmos. This included presuppositions such as the idea that a single God created the universe according to a specific and purposeful plan, and that all organisms— including humans—could (and should) be classified according to particular taxonomic criteria.

Science and technology continue to develop in Canada largely in response to environmental conditions and through increased contact with natural philosophers, engineers, and trade with other countries such as the United States. By the turn of the nineteenth century, science moves into the academy through the opening up of universities across Canada. Government at the federal and provincial levels begins to fund science and technology, and provide the organizational structures that help formalize distinct disciplines within science such as physics, botany, and geology. Both World Wars have a significant impact on what the Canadian government directs its funding and support towards (away from geology and towards solid-state physics, for instance). The wars combined with increased trade and so on introduce Big Science: science involving multiple groups of scientists in multiple laboratories within and outside Canada, funded increasingly not only by governments but also industry. Science and technology experience a number of government funding and investment peaks and troughs in the twentieth and twenty-first centuries (a peak in the aftermath of the Second World War compared with the current trough). Canadian funding for science and technology innovation remains modest compared to countries such as the United States, which, given the pre-eminence of science and technology as a driving force in nations' capacity-building, trade, and commercialization, is cause for concern.

This chapter then proceeded to outline the growing field of science and technology studies within sociology. Sociologists use their unique sociological imagination to study why and how science and technology practices work in particular ways. This style of analysis focuses on individual scientific practices in the context of much wider political, economic, social, and cultural commitments and practices. Using a sociological imagination also means paying special attention to the ways in which science and technology are integrated to such a degree as to become a salient part of the taken-for-granted fabric of our cultures. Society produces science and technology, but science and technology also produce society.

From here, we learned that modern science is both an idea and an ideal. The Enlightenment idea of modern science was centred on the presupposition and promise that science could derive true knowledge about the universe through rationality and empiricism exemplified in the scientific method. This idea accompanied the ideal of modern science to provide the *best* way of understanding our world, and structuring our societies according to that understanding.

Sociologists are particularly concerned to critique science and technology as grand narratives. A number of sociologists contest the very concept of modern science, arguing that it is predicated on an artificial and never-realized separation of humans from the rest of the universe. Moreover, sociologists argue, modern science never replaced religion as the only, or even primary, way of knowing the universe. Finally, sociologists argue science has never been *pure* in the sense that it has never practised according to objective, rational, and linear principles. Indeed, sociologists argue, science is social relations through and through. The remainder of this book considers this bold, and perhaps counterintuitive, claim. Chapters 2 and 3 focus on what *theories* sociologists have developed to analyze science and technology as social relations. Chapter 4 takes a closer look at the *methods* sociologists use to analyze science and technology as social relations. Chapter 5 explains *why* sociology considers science and technology as social relations, focusing especially on science as power. The book closes with a final chapter concerned with science and technology's relation to society. Rather than entirely determine, or be entirely determined by, society, this chapter argues science, technology, and society are entangled in a dialectical relationship, each shaping the other.

Key Terms

Interdisciplinarity   This term refers to the practice of different disciplines working together on the same research problem.

Clinical drug trials   Clinical researchers test new drugs through standard practices. The purpose of the trials is to ensure the safety and efficacy of the drug.

Great Chain of Being   This is a Christian and medieval conception of all living and nonliving matter on Earth in terms of a hierarchy.

Heliocentrism   This refers to the name given by Copernicus to describe his cosmology, or model of the sun at the centre of the universe.

Natural philosopher   This is the name given to people who studied nature and the physical properties of the universe. The term predates the modern term 'scientist'.

Protestant Reformation   This refers to the revolt beginning in the 1500s in Europe protesting the doctrines of the Catholic Church. The Reformation culminated in the ascendancy of Protestantism.

Enlightenment   This was a period of western history during which reason was understood to be the primary means through which to understand the universe and to conduct all human social, economic, political, and cultural relations.

Critical Thinking Questions

1. Are Canadians more ambivalent about the impact of science or religion on our society?

2. How confident are young Canadians that new discoveries in science and technology will solve pressing global problems such as global warming and/or world hunger?

3. What do you know about the history of science and technology amongst early Native and Inuit peoples of Canada? To what degree is Canadian science and technology multicultural?

4. Who was responsible for the atomic bomb: physicists, US Presidents, military personnel, the public, or the enemy?

Suggested Readings

W. Bauchspies, J. Croissant, and S. Restivo, *Science, Technology, and Society: A Sociological Approach*. Oxford: Blackwell Publishing, 2006. This is a comprehensive general introduction to science and technology studies.

D. Bell, *Science, Technology and Culture*. Milton Keynes: Open University Press, 2006. This general text provides an overview of the main theories relating science and technology to western culture.

S.H. Cutcliffe and C. Mitcham, *Visions of STS: Counterpoints in Science, Technology, and Society Studies*. New York: State University of New York Press, 2001. This edited collection contains chapters from ten leading scholars in Science and Technology Studies who argue from different perspectives for the overlap and integration of science, technology, and culture.

A. Ede and L.B. Cormack, *A History of Science in Society: From Philosophy to Utility*. Peterborough, Ontario: Broadview Press, 2004. This comprehensive book provides an excellent overview of the history of science in western and, to a lesser extent, eastern societies.

I. Hacking, *The Social Construction of What?* Cambridge, MA: Harvard University Press, 1999. This book, written by Canada's most well-known philosopher of science and technology, tackles the vexing and unresolved topic of social constructionism versus realism.

S. Harding, *Is Science Multi-cultural? Postcolonialisms, Feminisms, and Epistemologies*. Bloomington, Indianapolis: Indiana University Press, 1998. This book provides an historical and critical analysis of the myriad ways eastern science and women's participation in science has been consistently stolen, incorporated, elided, and then its origins erased from western science.

J. Law, 'On Sociology and STS', *The Sociological Review* 56, 4 (2008): 623–49. This article provides a useful overview of science and technology studies from a distinct sociological perspective.

S. Sismondo, *An Introduction to Science and Technology Studies*. Oxford: Blackwell Publishing, 2004. This book provides an overview of the main theoretical and methodological themes in science and technology studies.

Websites and Films

International Academy of the History of Science
www.aihs-iahs.org
This website provides the history of this international organization (which originated in France) as well as publications and prizes supported by the organization.

History of Science Society
www.hssonline.org
The History of Science Society is an association dedicated to studies in the history of science. It holds an annual conference, awards prizes to the best annual publication, and so on.

The Royal Society
http://trailblazing.royalsociety.org
This website, from the renowned (British) Royal Society, provides information about the history of scientific achievements in western societies.

Contributions of 20th Century Women to Physics
http://cwp.library.acla.edu
This website focuses on the diverse and important contributions of women to physics in the past century.

Nobel Foundation
http://nobelprize.org
The Nobel Foundation awards the most world-famous annual prizes for science disciplines.

The Complete Work of Charles Darwin Online
http://darwin-online.org.uk
This website provides excellent information about Charles Darwin's life, his research and publications, and the many critics and analyses of his theories.

Situating Science
www.situsci.ca/project-summary-clustering-humanistic-and-social-studies-science-canada

This website provides information about the 'Situating Science' cluster of researchers in Canada, currently funded by the Social Sciences and Humanities Research Council of Canada.

The Evolution of Eyes
www.youtube.com/watch?v=2ybWucMx4W8

This excellent short film provides a cogent explanation of how the human eye—purported by intelligent design proponents to be too complex to have evolved through natural selection—could have evolved through natural selection.

Planet Earth

This DVD series produced by the British Broadcasting Corporation presents life on planet Earth in all of its spectacular wonder. It presents science and technology as key means through which we approach a better understanding of life on Earth.

The Nature of Things with David Suzuki: Volume 1: Visions of the Future

Available in DVD format, the Canadian Broadcasting Corporation has produced this distinctly Canadian weekly program. It features Dr David Suzuki narrating some of the best episodes of the television series. It asks important questions about the future of life on Earth and our part in shaping the kind of future to which we may all look forward.

2 Science is Social Relations: Part I

Learning Objectives

In this chapter we learn:
- ⚙ Sociologists argue science is inherently social;
- ⚙ Sociologists propose different ways of understanding what science *is* and what science *does* in society;
- ⚙ Sociologists disagree as to whether some element (materiality) escapes the social character of science;
- ⚙ Sociologists disagree about whether or not we can (and should) take nonhuman entities (animals, trees, gravity) as part of what makes science social.

The Science Wars

In 1996 a physics professor at New York University published an article in the non-refereed academic journal *Social Text*. The article, 'Transgressing the Boundaries: Toward a Transgressive Hermeneutics of Quantum Gravity', ostensibly argued quantum theory supports postmodern arguments about the contingency of all meaning. Once the article was published, Alan Sokal revealed he wrote the article as a ruse to reveal social science and humanities' inappropriate use of science to make political claims. It became known as Sokal's Hoax, and it contributed to the Science Wars (Sokal, 2010; Lingua Franca, 2000; Bricmont and Sokal, 1999).

Sokal argued in favour of realism. He maintains:

> There *is* a real world; its properties are *not* merely social constructions; facts and evidence do matter. What sane person would contend otherwise? And yet, much contemporary academic theorizing consists precisely of attempts to blur these obvious truths. (1996: 51)

In response to Sokal's claims, a number of well-known scholars, including Bruno Latour, Margaret Lock, Isabelle Stengers, Donna Haraway, and others, offered counter-arguments that we might generally describe as social constructionist. For instance, Evelyn Fox Keller argued sociologists have turned to analyses of the social construction of science because 'truth and objectivity turn out to be vastly more problematic concepts than we used to think, and neither can be measured simply by the weight of scientific authority, nor even demonstrations of efficacy' (2000: 59). What Keller is arguing here is that it is not enough for scientists to just say that something is true, or even to demonstrate that objects work in particular ways. Such demonstrations do not reveal the inherently social processes of science. In sum, sociologists argue that: (1) science is inherently social, (2) scientists do not have objective access to reality, and (3) there are other ways outside of the scientific method to define reality.

Scientific Realism and Anti-realism

The Science Wars drew attention to at least two different and incompatible ways of understanding reality. Before proceeding with sociological theories about science and technology, it is useful to briefly explore some of the main terminology used within the philosophy of science because these concepts inform sociological debates about how to study science.

Scientific realism refers to the idea that a real world exists, and that science describes this real world (Leplin, 1984). As such, scientific claims can be either true or false. If a scientific theory accurately describes the world, it is true. If the theory does not describe the world, it is false. Scientific realism also maintains reality exists independently of any mind's perception of it.

Scientific realism has a number of well-known contemporary proponents, such as the Canadian philosopher of science, Ian Hacking (1983). Hacking describes how he came to identify his philosophy as scientific realism through an encounter with a physicist. Hacking was having a conversation about his friend's ongoing experiment to detect the existence of fractional electric charges, called quarks (ibid., 22). The complicated experiment consisted of suspending a tiny negatively charged droplet of oil between electrically charged plates. At some point in the historical development of this experiment, the tiny drop was changed to a big ball.

Hacking asked his friend how the charge on the ball was altered, to which his friend replied, 'Well, at that stage . . . we spray it with positrons to increase the charge or with electrons to decrease the charge' (ibid., 23). 'From that day forth' writes Hacking, 'I've been a scientific realist. *So far as I'm concerned, if you can spray them then they are real*' (1983: 23). This statement—'If you can spray them, then they are real'—is one of the best-known contemporary by-lines of scientific realism.

Plato's *Dialogues* defined this debate as a battle between the gods and the Earth giants (Goldman, 2006). Plato's gods were on the side of certain knowledge, truth, reason, and reality. That is, knowledge as Certain, Universal, Necessary, and True. The work of the philosopher and contemporary scientist is, paraphrasing Einstein, to 'lift a corner of the veil'. For Plato, this kind of knowledge is exemplified by mathematics because it follows a strictly deductive form of reasoning. Against these gods of certain knowledge, Plato pitted the Earth giants (Sophists), who understood knowledge as a certain type of belief.

Science is replete with larger-than-life figures firmly on the side of Plato's gods. Galileo Galilei's equation for a freely falling body, $s=1/2gt2$, where the distance covered by a freely falling body and the time is directly proportional to the time squared, exemplified the orderly nature of reality revealed through mathematics. Galileo's telescope observations of the moons of Jupiter and his numerous *gedankens* were directed at discovering true knowledge of reality.

Isaac Newton, born the year Galileo died, is, next to Plato, most securely associated with scientific realism. As Alexander Pope once remarked 'And God said, "Let Newton be", and there was light'. Newton's equation for gravity began with a set of principles from which could be derived universal, certain, necessary, and true knowledge. For instance, Newton defined space and time as uniform and infinite in all directions. We return to Newtonian physics (also referred to as solid state or classical physics) in Chapter 3.

Scientific anti-realism, by contrast, argues that objective reality does not exist (Dummett, 1978). Whereas realism argues unobservables (quarks, DNA, molecules—things that cannot be detected with human senses alone) exist, anti-realism argues they do not exist (Hacking, 1999). During Greek antiquity, the Sophists argued knowledge is a form of belief, and a position of skepticism concerning the impossibility of true knowledge has filtered down through philosophy ever since. George Berkeley, David Hume, and John Locke championed empiricism, the theory that all knowledge derives from our senses that formulate our experiences and knowledge of the world onto the mind. We achieve knowledge about nature by applying reason to primary sensations such as the size and shape of entities. Our knowledge is a derived one, based on the accumulation of experiences, and thus can never be certain, universal, necessary, and true. (This is also the basis of phenomenology, which is an approach that focuses on how humans make sense of the world through their apprehension and experience of phenomena).

The German philosopher Immanuel Kant attempted to bridge Newtonian physics with anti-realist skepticism, found, for instance, in David Hume's empiricism. At the core of Kant's philosophy is the idea that while absolutely true knowledge might be impossible to ascertain, we may come as close as possible (as close as we would ever practically need) by recognizing that experience is constructed by the mind, and the mind conforms to certain universal, uniform, and consistent rules. Time and space are examples of such rules, and they are hard-wired (Kant called this intuition) in our brains. Intuition produces our judgment of the truth of knowledge prior to experience. In other words, we do not acquire our knowledge through experience per se, because our experience is always already created through our intuition. Kant argued that we are able to arrive at knowledge through reason, but we cannot ascertain absolute knowledge that is independent of our response to it. Cavell wryly summarizes Kant's approach:

> To settle with skepticism . . . to assure us that we do know the existence of the world, or rather, that what we understand as knowledge is of the world, the price Kant asks us to pay is to cede any claim to know the thing in itself, to grant that human knowledge is not of things as they are in themselves. You don't—do you?—have to be a romantic to feel sometimes about that settlement: Thanks for nothing. (1988: 31)

Skepticism towards absolute truth has at its disposal a social scientific critique focused on the premise that all natural sciences depend upon one or more initial propositions, and that these propositions are born of the *social enterprise* that is science itself. In the next section, we take a closer look at what is referred to as the social construction of scientific facts.

Socially Constructed Matters of Fact

Sociology students are probably familiar with the term social constructionism. It serves as an umbrella concept that describes ways in which humans generate meanings about the world. Postcolonial studies focus on the ways in which indigenous and settler identities are made meaningful through historical and contemporary discourses about nationhood. Feminist studies focus on ways in which sex and gender are attributed particular meanings that result in differential treatment of girls and women in all cultures. These analyses share a commitment to the central idea of social constructionism: that social structures are not reified by nature or a natural order of things, but are, rather, effects of conditions and forces that benefit some groups at the expense of others and that could be otherwise.

The social construction of scientific facts broadly focuses on science as a set of explicit and implicit practices. Sociologists argue real objects exist. But, at the same time, they argue, science is an inherently social practice. As this book will show, science involves communication amongst scientists; dialogue and debate (agreement and disagreement); styles of reasoning and experimentation; criteria for qualifying as a scientist; the construction and interpretation of data; the construction, use, and interpretation of technology; and so on. More generally, science takes place within a cultural and political milieu, which sets the parameters of what is known and what can be known at any given time and in any given place. These practices are necessarily dynamic in that not only *what* we know, but *how* we know what we know, change over time.

Let's illustrate the companion ideas that what we know and how we know changes over time with two examples. The first example brings us back to the Renaissance, at a time when philosophers, astronomers, and poets gazed skyward and attempted to understand our place within the universe. The Copernican Revolution was, as we saw in Chapter 1, a literal and metaphorical shift in how people viewed the cosmos. The Catholic Church considered Earth to be the centre of the universe, with the stars and the rest of the planets revolving around it. The Church maintained this view because it supposed God would have placed humanity at the centre of the universe. Ptolemy followed the philosophy of the Church by depicting the universe as a set of nested spheres, with Earth at the centre. Galileo Galilei, also an astronomer, physicist, philosopher, and mathematician, disagreed. He defended the astronomical theory of his predecessor, the sixteenth-century astronomer Copernicus, who depicted the sun at the centre of the universe. Galileo invented and refined telescopes that allowed him to see planets and stars with more precision than previous astronomers. But he also used a carefully constructed argument. And this argument, however rationally guised, required quite a leap of faith:

> Copernicus's theory requires us to believe contrary to all experience that the Earth is rotating on its axis at approximately 1,000 miles per hour. The Earth's circumference is about 24,000 miles, so in a 24 hour period, we have to cover 24,000 miles. If you throw a ball up in the air, how come it isn't blown backwards? If a bird takes off at the equator, why doesn't a 1,000 mile-an-hour wind blow it backwards? Why don't we see any evidence of this motion? If the Earth circles the Sun, then it means that in June and December, the Earth is on opposite sides of the Sun. Then how do two stars that are lined up in a row in June still line up in December? (Goldman, 2006: 44–5)

This leap of faith involves what sociologists of science call representationalism, or representative realism. Representationalism is basically the idea that humans cannot know the external world as it is: we only know the world around us through the interpretation of our perceptions. In this way, Copernicus and Galileo were requiring people to believe things counterintuitive to their perception of the world. We perceive our world to stand still: we even (erroneously) say that the sun rises and sets. The Copernican system states that Earth moves and that it is Earth, as it were, that rises and sets.

We might say that Galileo was ultimately right: scientists have long since provided compelling evidence that Earth does revolve around the sun. But it's more complicated than this. History being on the side of the winners, we take Newton's vindication of the Copernican system, which he substantiated using Johannes Kepler's ideas based on Tycho Brahe's research (which Kepler purportedly stole after Brahe's death). Yet Brahe, a Danish astronomer, came up with a theory of the heavens in which Earth rotates on its axis but is stationary and the sun orbits Earth, Mercury, and Venus, the inner planets orbit the sun as it orbits Earth, and Mars, Jupiter, and Saturn orbit both the sun and

Earth. Moreover, Galileo did not (nor could he have) come up with an experiment that would have disproved Brahe's system while substantiating Copernicus's. Although Galileo knew of both Brahe's theory and Kepler's idea that the planets orbit the sun in ellipses, he ignored both. Galileo's *Dialogue on the Two Great World Systems* (1632) is a fictitious conversation between Ptolemy and Copernicus, with Copernicus in the clear lead. Galileo deleted some information, emphasized other ideas, and based his work on a combination of theoretical speculation, creative *gedankens* (see Chapter 3), and witty argument (Goldman, 2006).

Since we now have evidence to suggest that planets orbit in ellipses but not at uniform speeds, we can reasonably say that Galileo and Newton were wrong even though they all thought, and we maintain today, that they engendered true knowledge. So the scientific enterprise is more complicated than we typically allow. Just as we now agree that a great deal of the science during this period, including that of Galileo and Newton, was erroneous, it is reasonable to assume that some things we now hold to be scientifically substantiated will, 400 years from now, be considered incorrect. Galileo is by no means the only example. Thaddeus Trenn, for example, analyzed the way in which scientists 'prematurely falsified' the theory that thoruranium (a type of uranium) is the extinct 'parent' of thorium (1978), and the scientific acceptance of aggregate recoil theory that was later acknowledged to be incorrect (1980).

The second example comes from an enormously popular sociology of science book by Steven Shapin and Simon Schaffer, entitled *Leviathan and the Air Pump: Hobbes, Boyle and the Experimental Life* (1985). The book concerns the invention of the air-pump vacuum. Robert Boyle (a gentleman scientist of titled background and considerable means, focused on physics and chemistry) and Robert Hooke (a wealthy natural philosopher) developed a series of experiments on the properties of air using a pump invented in the seventeenth century. These experiments were widely acclaimed and caught the critical attention of Thomas Hobbes (a philosopher of modest means). A critical debate ensued, involving the Royal Society, natural philosophers, and others.

The air pump consisted of a suction pump attached to a replaceable glass bulb that could be replaced. When the pump worked, it created what we now think of as a vacuum in the glass bulb. At the time, however, what exactly this space consisted of was a matter of debate. Priests representing the Church's position argued that when the air pump pumped out air, it left an incorporeal substance. The theory that there is no such thing as empty space is called plenism, and priests argued the immortal soul is what's left after matter is removed. While Boyle wanted to stay out of this debate (we learn more about Boyle's philosophy next), Hobbes was squarely on the side of the state, arguing that dividing people's allegiance between the Church and Crown created civil unrest and the risk of civil war. Using the air pump to make a plenist argument was, for Hobbes, a way for priests to 'usurp power' from the King (ibid., 96).

Boyle was predominantly concerned to define matters of fact. He performed two main experiments using the air pump to demonstrate how we should arrive at matters of fact. In the first experiment, a Torricellian apparatus (which consists of liquid in a tube) is placed in the bulb. The liquid in the tube falls, but not quite to the level of the liquid in the dish placed at the base of the inverted tube. Boyle interpreted this experiment to mean that as air was evacuated from the bulb the tension was no longer acting on the liquid. The liquid doesn't completely fall, according to Boyle, because of a small amount of air remaining in the bulb due to leakage. For Boyle, to say that when air is sucked out of the bulb, the level of the liquid in the inverted tube fell, is a matter of fact. In the second experiment, Boyle placed two smooth glass discs of equal weight and size into the bulb, predicting they would separate when the air pressure was removed (i.e., that the air pressure would keep the two glass balls together). Interestingly, when Boyle performed this experiment, the glass balls did not separate, and Boyle accounted for this result by reasoning that he could not reduce the air pressure sufficiently, and that the air pump leaked air.

Boyle essentially created a series of arguments about the experimental method (see Chapter 4). He argued that matters of fact should be derived from carefully constructed and conducted experiments that other people could replicate either directly by constructing their own pumps according to meticulously detailed laboratory notes (instructions), or virtually by imagining the air pump operation through detailed description. Boyle's matters of fact, then, were based on hands-on experiments that produced probabilities of things happening rather than absolute, certain truth. Boyle did not consider his rejection of absolute certainty to be 'a regrettable retreat from more ambitious goals' but rather something to be 'celebrated as a wise rejection of a failed project' (ibid., 24). According to Shapin and Shaffer, Boyle deployed three 'technologies' to produce knowledge (matters of fact): 'a *material technology* embedded in the construction and operation of the air pump; a *literary technology* by means of which the phenomena produced by the pump were made known to those who were not direct witnesses; and a *social technology* that incorporated the conventions that experimental philosophers should use in dealing with each other and considering knowledge-claims' (ibid., 25). Matters of fact, for Boyle, could be derived without universal agreement.

Hobbes vehemently disagreed. For Hobbes, matters of fact are derived from logic, exemplified most clearly in geometry. This logic is necessarily universal, entirely replicable (everyone derives the same answer by performing the same geometric equation), and is not based on messy experiments that produce contradictory results that not everyone can either observe for themselves, or agree upon. Hobbes further argued that experiments could not reveal causes because it is impossible to separate facts of nature from any given artifact produced by the machine/experiment itself. As Goldman (2006) points out, it is one thing to agree that an X-ray is a valid and accurate depiction of a bone; it is another to agree that a particular shadow represents an aneurysm. To agree on the latter requires specialized training with the measuring instrument, which necessarily introduces a social dimension (classification scheme and so on) into the determination of truth.

While Hobbes is usually regarded as the loser in the air pump debate, he did in fact have a rather good point. Since machines necessarily already embody the theories used to build them (X-ray machines are built to provide X-ray images and air pumps were invented to create vacuums), how can we with any certainty differentiate between the measuring instrument and the theory? In other words, 'the expectation of what the machine is going to show you is already there. How can you use that as a confirmation of the expectation?' (Goldman, 2006: 78). And how can we know, with certainty, that the data collected from measuring instruments are a faithful, exact, and consistent reflection of the object of study?

Moreover, Boyle was introducing a kind of citizen science (see Chapter 6) by suggesting that matters of fact (facts, in this case, about air vacuums) could be arrived at by the consensus of a number of people witnessing the air pump experiment. The vision may have been to get an air pump to every school in Britain, but the reality was that air pumps were difficult to obtain, complicated to assemble, and finicky to handle. Oxford University had an air pump, as did a laboratory in London and one in Germany. Witnessing air pump experiments, then, needed to be virtual, and Boyle described the ways in which the scientist would need to detail his (there was absolutely no suggestion that a scientist could be female) procedures and experiments so that everyone could agree on their reliability and validity. This considerably whittled down the number of people who would actually see the air pump working. Boyle also had a lot to say about what kind of person was an appropriate experimental witness. Not just anyone could walk off the street. He had to be known and respected in the polite society of his culture, in order to be a 'modest witness' to the air pump experiment. In short, Boyle argued for the *opposite* of open access to scientific work. He introduced what is known as the scientific expert (see Chapter 6). And this necessarily introduces a further social dimension, as experts must be educated both in the subject matter of any given field of expertise and also that field's culture. And the idea that scientists do their work within a particular culture brings us to a consideration of science as a collective social enterprise.

Why Science and Technology are Social

Sociologists, as we have learned, consider science to be an inherently social enterprise. What sociologists mean is that scientific research depends upon a number of human-created concepts such as expertise, rationality, paradigms, and so on. Now that we have a preliminary understanding of the differences between realism and social constructionism, the remainder of this chapter considers the main theories sociologists and philosophers of science have developed to analyze science. One important point to keep in mind as we consider each of these theories is that for the most part sociologists endorse the claim that objects exist in the world (I have never met a sociologist who doesn't believe in gravity or rocks). What sociologists argue is that science and technology are necessarily social because they are human endeavours to better understand, interpret, and work with entities in the world.

The Political Economy of Science and Technology

Karl Marx's theory of capitalism is based upon the differential relation labour has to the forces of production. The fundamental struggles between social classes, Marx argued, is based on the domination of the bourgeois (middle and upper classes) over the proletariat (working class labourers) through the surplus value (i.e., unpaid labour) created when the bourgeois do not pay the proletariat what their labour is worth. The proletariat's labour changes throughout history because of new developments in science and technology (printing press, loom, computer, and so on). The question is how Marx viewed these material forces of production. A number of sociologists argue Marx considered science and technology—and the entities scientists and engineers create and manipulate—to be both the central force of production in modern capitalism and independent of the social relations defining the political economy of society (Aronowitz, 1988). Moreover, Marx argued these productive forces determine the relations of production. That is 'although the appearance of new scientific theories or technologies depend upon social conditions, the content of these discoveries are in no way dependent upon social relations' (ibid., 37).

In its more contemporary iterations, Marxist theory has developed a critique of the ways in which bourgeois social relations affect the relation humans have to nonhuman entities, and control the nature of scientific discovery. It argues the capitalist bourgeois worldview, which includes the axiom that all objects can be manipulated for human ends, shows that science and technology are not neutral enterprises but part of the political economy of society. Moreover, Marxist theory argues capitalism is predicated on the development of science and technology because increasing capital requires new ways of increasing production. As Engels wrote, 'You must not separate agriculture and also technology from political economy. Crop rotation, artificial fertilizer, steamengine and power loom are inseparable from capitalist production . . .' (Marx and Engels in Aronowitz, 1988: 45). As such, science and technology are a social power within society. Further, as Aronowitz (1988) argues, each generation tends to forget the labour processes that create the forces of production as it incorporates the products of science and technology into their normal regular environment. For instance, people tend to see money and commodities (radio, iPhone, bicycle) as 'independent forces rather than as symbols of stored-up labor. . .or as the product of rationalized labor process[es]' (ibid., 47). We return to this discussion in Chapter 5 where the relationship between political economy, the interests of the elite, and science and technology are brought to the fore.

Thought Collectives, Paradigms, and Normal Science

Ludwik Fleck, a Polish doctor and biologist, developed a series of procedures that led him to invent a vaccine for typhus. During the Second World War the Nazi party arrested Fleck and his family, and forced him to work at Auschwitz and Buchenwald concentration camps. The Nazis wanted Fleck to develop a way to diagnose syphilis and other illnesses through a serotological method (tests using blood serum). In 1935, Fleck published a book entitled *Genesis and Development of a Scientific Fact* (1935/1979). At the time, this slim book attracted little attention. It wasn't until the late 1970s and early 1980s that sociologists began to take this book seriously.

The book examines the shift from understanding syphilis as a moral illness to perceiving it as a bacterial disease. Fleck shows how two different approaches to syphilis developed side by side: (1) syphilis as a 'carnal scourge' reflecting moral vice, and (2) syphilis as a disease. Fleck's point here is that variable definitions produce variable conclusions. Fleck distinguishes between experiments and experience: experiments derive simple answers from simple questions, whereas 'experience must be understood as a complex state of intellectual training based upon the interaction involving the knower, that which he [sic] already knows, and that which he has yet to learn' (ibid., 10).

Fleck argues science should not be understood as a step-by-step progress—adding fact upon fact—towards the truth of any phenomenon. Rather, scientific facts are constructed events that begin with a scientist's training. Training, argues Fleck, means associating one's self with a particular thought collective. Thought collectives essentially *constrain* observations, experiments, and experience into comprehensible information. Recognizing a fact, then, means misinterpreting or denying other facts: 'to recognize a certain relation, many another relation must be misunderstood, denied, or overlooked' (ibid., 30). Facts are related to thought collectives in three ways:

(1) [E]very fact must be in line with the intellectual interests of its thought collective, since resistance is possible only where there is striving toward a goal. Facts in aesthetics or in jurisprudence are thus rarely facts for science . . . (2) The resistance must be effective within the thought collective. It must be brought home to each member as both a thought constraint and a form to be directly experienced . . . (3) The fact must be expressed in the style of the thought collective. (ibid., 101–2)

> **BOX 2.1** ❈ **THE WASSERMANN TEST**
>
> The Wassermann test is named after August von Wassermann, a bacteriologist who invented a way to detect syphilis in which an antigen is introduced to a sample of extracted blood of cerebrospinal fluid. The problem with this test is that it produces false positives and false negatives. The reaction is not specific to syphilis and it will produce reactions to other diseases, including malaria and tuberculosis. Moreover, some people infected with syphilis show no reaction, and other people cured of syphilis continue to show a reaction.
>
> In *Genesis and Development of a Scientific Fact* (1935), Fleck points out that despite the fact that the test is unreliable and that it regularly produces unexpected results, it is nevertheless considered an important medical aid, has been the subject of many journal articles and reports, and is subject to official regulation. He argues that the test 'works' only within a particular thought collective that essentially formulates knowledge in such a way that excludes different formulations. Struggles to get the test to work, and then to be accepted as standard practice, rely on a particular concept of infectious disease, immunity, study replication, and so on— all of which have been challenged in the results of the test itself.

Fleck isn't arguing that thought collectives are bad or negative; rather, they are necessary if scientists are to organize and make sense of the vast quantity and diversity of observations we can make about the world. Facts are not arbitrary either, because they are a function of a thought collective (ibid., 156). Indeed, what Fleck calls 'thought styles' formulate concepts. Thought collectives, then, produce scientific facts, and are a 'prerequisite to any thought at all' (Kuhn in Fleck, 1979: xi). In other words, 'evidence conforms to conceptions just as often as conceptions conform to evidence' (1979: 28). Facts, then, 'are not objectively given' as the public often thinks, 'but collectively created' (ibid., 157). And as thought collectives change over time, so do our facts about the world. In other words, Fleck argued that science is social and dynamic.

In 1962, Thomas Kuhn published a book entitled *The Structure of Scientific Revolutions*. This short book has had a strong influence on the sociology of science, and is still frequently referred to today by sociologists. Kuhn was a Harvard-educated American physicist. Much like Fleck, Kuhn argued science should not be thought of as a steady progress towards an ultimate and independent truth about our world. Instead, all scientific research takes place within a set of shared paradigms, rules, procedures, and standards.

And like Fleck, Kuhn begins with an argument about pedagogy. Kuhn argues the textbooks from which trainees (undergraduate students, for instance) are taught are exemplars of the 'rigorous and rigid' education that

---

BOX 2.2 ❖ WHAT IS A PARADIGM? WHAT IS A PARADIGM SHIFT?

In its common usage, the word 'paradigm' means an accepted model or pattern.

Thomas Kuhn uses the word in a somewhat different sense, to mean the shared commitment that enables a particular model to be accepted within the scientific community. Put another way, Kuhn defines a paradigm as a shared commitment to the same rules and standards for scientific practice (1962: 11). Kuhn argues that whereas students in the humanities and social sciences are exposed to competing paradigms, science only operates insofar as there is one paradigm. For this reason, a paradigm shift constitutes a scientific revolution. A shift rarely occurs in science because it requires a change in the shared commitment to the underlying assumptions that enable science to work. Copernicus's work, as we saw in Chapter 1, led to a paradigm shift within science from an Earth-centred to a sun-centred universe. Quantum theory (see Chapter 3) led to a paradigm shift concerning how matter works.

---

scientists undertake (ibid., 5). This education does much to explain *how* science works and the *directions* it takes. When students join a particular field, they are learning a set of facts that have been produced within that field, through an inherently social process. When they join laboratories or research groups and begin to conduct science, they join colleagues schooled in the same paradigms: the same rules, procedures, standards, and ways of looking at the world. This shared set of assumptions and practices is a prerequisite for accomplishing science.

Kuhn introduces the concept of normal science to describe the continuation of a particular research tradition. Normal science is the bread and butter of the scientific enterprise. It consists of all the everyday practices that scientists engage in. Scientists perform these tasks—coming up with hypotheses, building equipment, performing experiments, working out equations, gathering data, analyzing data, writing up results, writing grant proposals, presenting at scientific conferences, and so on—within a shared set of assumptions. Therefore, normal science means 'research firmly based upon one or more past scientific achievements, achievements that some particular scientific community acknowledges for a time as supplying the foundation for its further practice' (ibid., 10). Normal science is conservative: it conserves, it is rooted in the past, it is resilient, and it defines good science (and scientists) in terms of conformity to the accepted paradigm.

A very important aspect of normal science is that it 'often suppresses fundamental novelties because they are necessarily subversive of its basic commitments' (ibid., 5). As such, according to Kuhn, most people who are not directly involved in science do not realize normal science is mainly a mopping-up exercise (Hacking, 1983). That is, the bulk of science is devoted to making facts fit the particular scientific paradigm within which the scientists operate at any given time. This insight runs counter to the way in which many non-scientists understand science. A common public perception of science is that it makes progressive discoveries. According to Kuhn, science works by doing the exact opposite.

Mopping-up operations are:

> an attempt to force nature into the preformed and relatively inflexible box that the paradigm supplies. No part of the aim of normal science is to call forth new sorts of phenomena; indeed those that will not fit the box are often not seen at all. Nor do scientists normally aim to invent new theories, and they are often intolerant of those invented by others. Instead, normal-scientific research is directed to the articulation of those phenomena and theories that the paradigm already supplies. (Kuhn in Fleck, 1979: 24)

Kuhn argues normal science follows a particular process. Science, says Kuhn, begins when scientists detect an anomaly in a particular theory. That is, when scientists find something that doesn't seem to fit with the paradigm, they experience it as an anomaly. They proceed to explore the anomaly with a view to eliminating it. Scientists can eliminate the anomaly by explaining it within the paradigm. If this can't be done, scientists may choose to defer the explanation to a later date when they have the appropriate equipment (which may not yet have been invented).

On rare occasions, according to Kuhn, the anomaly cannot be explained or deferred. Sometimes, indeed, the anomaly begins to spread because more and more scientists find the anomaly to occur in their own research results. Only now does normal science begin to falter, and only when there is an alternate paradigm to replace the current one. At these rare times, science faces a crisis, which is resolved when the current paradigm shifts to a new paradigm. In these cases, Kuhn describes these shifts as revolutions in science.

We might think, for example, of the crisis in physics prompted by quantum theory (see Chapter 3). Until then, Newtonian physics (or classical mechanics, as it is also called), with its emphasis on solid states, had been the dominant paradigm. When anomalies were discovered, such as the now-famous Wave-Particle Duality Paradox, they were defined as such—small *apparent* exceptions to the rules that would be explained in time so as to be incorporated into the paradigm. Only when more exceptions started cropping up at the quantum level did physicists begin to seriously turn their attention to the implications of these growing numbers of exceptions for the paradigm as a whole. Even when the paradigm shifted toward quantum theory, eminent physicists such as Albert Einstein remained openly skeptical and resisted the paradigm shift.

All Science is Social

The Strong Programme (called the Sociology of Scientific Knowledge, or SSK) is most closely associated with a number of social scientists based at Edinburgh University in the 1970s, including David Bloor and Barry Barnes, as well as social scientists located in other places, such as Harry Collins at Bath University. SSK builds upon many of Fleck and Kuhn's insights concerning science as a social enterprise based upon structures and processes within a given paradigm. SSK was originally concerned that social scientists tended to accept scientists' demarcation of good versus junk or pseudo-science—quantum theory versus paranormal activity for example—rather than seeing both fields of study through a sociological lens.

To view science through a sociological lens, Bloor developed the symmetry postulate that argues all ideas (whether rational, irrational, true, or false) are entirely social as opposed to the Kantian idea that true (rational) beliefs are based on reality while false (irrational) beliefs are based on societal influences (Collins and Yearley, 1992: 304). For Kant, the true belief that Earth revolves around the sun is based on rational science (employing the scientific method) whereas the false belief that the stars and planets revolve around Earth was based on Christian Church influences. In contrast, SSK holds both true and false beliefs (as in the example above) on equal footing because the correspondence of beliefs to reality is itself based on subjective assertion, imputation or acceptance.

Thompson provides a flavour of this perspective:

> It seems ironic that human experiences known by artists and saints and yogis in different cultures over the millennia, and repeated over and over again in quite different situations, are dismissed as superstition and illusion, but an elementary particle that only exists as a nanosecond impulse on a screen seen only by a handful of high priests at CERN at a cost greater than the construction bill for the Great Pyramids is considered to be 'scientifically real'. Elementary particles are no more real than angels or garden dwarves; they are, in Varela's words, 'brought forth'. Elementary particles are brought forth by linear or ring accelerators, just as angels or bodhisattvas are brought forth by meditation. Physics . . . is a language. (1991: 20)

The Strong Programme argues society and nature are not fundamentally separate because society is part of nature, and knowledge itself is a natural phenomenon:

> We can assume that observation will always enable us to uncover a reality, which is more complicated than we can assimilate into our current conceptual schemes and theoretical systems. Experience and practical involvement with the world will endlessly generate anomaly. *Nature will always have to be filtered, simplified,*

selectively sampled, and cleverly interpreted to bring it within our grasp. It is because complexity must be reduced to relative simplicity that different ways of representing nature are always possible. *How we simplify it, how we cho[o]se to make approximations and selections, is not dictated by (non-social) nature itself. These processes, which are collective achievements, must ultimately be referred to properties of the knowing subject.* This is where the sociologist comes into the picture. (Bloor, 1999: 90, emphasis added)

For SSK, nature is more complex than observers' attempts to explain it: we come to understand parts of nature through an endless refining process of observation and interpretation. It is this refining process that sociologists seek to analyze, by ascertaining detailed knowledge about what scientists are responding to—their stimuli—whether neutrinos, genes, or bacteria. Again, nature affects our beliefs about how nature is experienced, but it does not causally explain how it is then described.

As such, Bloor describes SSK as relativist in the sense that it is concerned to evaluate theories and beliefs in terms of their credibility within science, which necessarily engages with science as a social enterprise, consisting of the context in which any given theories or beliefs are found (including classification, rules, principles, and so on). Not surprisingly then, Bloor does not characterize science as a zero-sum game in which the more we know as fact, the less it is based on our interpretations of the world, because for Bloor, 'all knowledge always depends on society' and is thus necessarily provisional (ibid., 110).

Adherents argue SSK provides a method that solves the problem of relying upon scientific descriptions and explanations of nonhuman entities. However, in practice SSK aims to 'explain shared beliefs about nature' and nature seems to remain effectively silent (ibid., 87). What Collins and Yearley identify as SSK's strength, to '. . . show that the apparent individual power of the natural world is granted by human beings in social negotiation', is considered a weakness by a number of social scientists who take a different approach to the study of science (Collins and Yearley, 1992: 310).

Everything is Social Relations, but Not Everything is Human

Actor Network Theory (ANT) is most closely associated with the work of Bruno Latour, John Law, and Michel Callon. In contrast to the Strong Programme, ANT seeks to place things-in-themselves at the analytic centre:

Things-in-themselves? But they're fine, thank you very much. And how are you? You complain about things that have not been honored by your vision? You feel that these things are lacking the illumination of your consciousness? But if you missed the galloping freedom of the zebras in the savannah this morning, then so much the worse for you; the zebras will not be sorry that you were not there, and in any case you would have tamed them, killed, photographed, or studied them. *Things in themselves lack nothing,* just as Africa did not lack whites before their arrival. (Latour, 1988: 193, emphasis added)

Latour charges that western society is predicated on an artificial separation between nature and culture in which knowledge about nature's order of things is *sui generis* or self referential (that is, knowledge entirely contained within that defined as sociality): a split that never actually happened, leading Latour to conclude that *We Have Never Been Modern* (1993). Latour's main point is that the separation between culture and nature is an artificial, philosophical enterprise: 'forces cannot be divided into the "human" and "non-human" . . . It is not a question of *nature . . . Natures* mingle with one another and with "us" so thoroughly we cannot hope to separate them and discover clear, unique origins to their powers' (ibid., 205–6). Latour deploys the term 'collective' to describe human–nonhuman associations and relegates society to one particular part of the collective, 'the divide invented by the social sciences' (1993: 4):

> . . . [T]here are two and only two known and fixed repertoires of agencies which are stocked at the two extremities—brute material objects, on the one hand, and intentional social human subjects, on the other. Every other entity—gravitational waves, scallops, inscriptions, or door closers, to name a few—will be read as a combination or mixture of these two pure repertoires. (Callon and Latour, 1992: 350)

According to ANT, the world is made up of actants, defined as anything that relates to other actants: 'atoms and molecules are actants, as are children, raindrops, bullet trains, politicians, and numerals . . . An atom is no more real than Deutsche Bank or the 1976 Winter Olympics, even if one is likely to endure much longer than the others' (Harman, 2009: 12).

Actants are entirely concrete and in a specific place in the world, and with specific relations with other actants at any given moment. Actants are not powerful through some inner essence, but rather attain force through assembling allies. This process is temporal and always subject to disintegration: 'it is never an actant in naked purity that possesses force, but only the actant involved in its ramshackle associations with others, all of which collapse if these associations are not lovingly or brutally maintained', in what Latour refers to as a 'tiered array of weaknesses' in which the 'winner has stronger alliances' (ibid., 61). In short, realism, for ANT, is *resistance* (ibid., 20, 29):

> What makes the atom more real [than a ghost] is that it has more allies, and these allies stretch well beyond humans. Experiments testify to the atom's existence; instruments stabilize it and make it indirectly visible; generations of children learn

of it and pass the word along; Brownian motion shows that particles of water are moved by it. The ghost, by contrast, has only a paltry number of allies bearing witness to its reality. But the atom's allies may one day desert it too. (ibid., 144)

The problem with SSK, for Latour, is that it does not allow that nonhuman entities themselves affect our understandings (beliefs, theories) about them. Contrasting SSK's advancement of symmetry with regard to valid and invalid science, Woolgar refers to ANT's 'radical symmetry with regard to agency' (1992: 335). Indeed ANT proponents reverse what they see as SSK's desire to 'strip science of its extravagant claim to authority' by asserting 'nature settles controversies' (Callon and Latour, 1992: 346). Put another way, in SSK 'the belief system has to register the world without the world introducing any significant difference, apart from its mute presence and insistence' (Latour, 1999: 118). For ANT:

Nonhumans are party to all our disputes, but instead of being those closed, frozen, and estranged things-in-themselves whose part has been either exaggerated or downplayed, they are actants—open or closed, active or passive, wild or domesticated, far away or near, depending on the result of the interactions. When they enter the scene they are endowed with all the nonhuman powers that rationalists like them to have, as well as the warmth and uncertainty that social realists recognize in humans. (Callon and Latour, 1992: 355)

In other words, reality for Latour is social relations. Graham Harman summarizes ANT's distinct approach:

[ANT] has nothing to do with old-fashioned realism, since it places physical mass on the same level as puppet shows and courtroom hearings. It has nothing to do with social constructionism: after all, it is not limited to human society, which is pounded by the demands of nonhuman actants as if by waves of the ocean. . . . It is not phenomenology, because an electric drill or vein of silver are not appearances for human consciousness, but actants that undermine whatever humans encounter of them. (2009: 29)

What is most promising in Latour's theory is that it explicitly acknowledges that actants, and relations between actants, need not have anything to do with humans. That is, objects do not require human mediation in order to act: 'What makes a hybrid a hybrid is not its combination of nature and human civilization, but more generally its fusion of substance and network' (ibid., 7).

A criticism of ANT's account is that objects do not seem to have any status outside of the events in which they relate: alliances seem not only to articulate objects but to create them as well. Another limitation is that while ANT champions an approach that takes actants seriously in their own right, its empiricism resembles the kind familiar to SSK; that is, studies that focus much more on definably social actors (paradigms and politics, for instance) than material objects. In practice, when empirical studies emanating from SSK and ANT are compared, they appear similar in their focus on sociologically familiar objects of study such as norms, paradigms, and data dissemination rituals. Even laboratory devices are described and interpreted more in terms of their human invention and attribution of meaning than in their material composition and operation: *scientists* in action as it were. As Harman observes:

> . . . [Latour's] examples are drawn from the human realm, not from general cosmology. . . . With a bit of work, it is not difficult to see why all objects that enter human awareness must be hybrids, why the ozone hole or dolphins or rivers cannot be viewed as pure pieces of nature aloof from any hybridizing networks. The harder cases involve those distant objects in which human awareness is currently not a factor at all. Where are the hybrids in distant galaxies? (ibid., 7)

So how might social scientists study actants in all of their messy relations, while also acknowledging that nonhuman entities have some sort of agency in their own right, and are certainly independent of human interactions with them? For an intriguing answer, we turn to a science studies scholar who says we need to explore the mangle of scientific practice.

Science and Technology Mangles Humans and Other Entities

Our final theory draws from both the Strong Programme and Actor Network Theory. Andrew Pickering, a physicist turned science studies scholar, was concerned to develop a theory for understanding scientific practice that understood science as an inherently social enterprise *and* objects as having agency in their own right. In two books, *Constructing Quarks: A Sociological History of Particle Physics* (1984) and *The Mangle of Practice: Time, Agency, and Science* (1995), as well as numerous articles Pickering offers what he calls a pragmatic realism to describe a way of understanding the relationship between the world and scientific knowledge (1995: 183–5). Pickering argues matter and scientific knowledge come into contact with each other through a mangle (or in its verb form, through mangling).

As Pickering sees it, humans do not control the world or direct the entire flow of agency. Hurricane Katrina, asteroids bombarding Earth, and snowdrifts are all examples of the agency of the world. Indeed, 'much of everyday life', observes Pickering, is devoted to 'coping with material agency, agency that comes at us from outside the human realm and that cannot be reduced to anything within that realm' (ibid., 6). Pickering advances science as performative but not in the sense social constructionists tend to mean, whereby culture entirely directs phenomena. For Pickering, scientists interact with matter over time to produce things (quarks, for example) and knowledge (quantum theory, for example). Matter has its own agency. For this reason, Pickering is critical of SSK's disinterest in taking nonhuman agency into account.

Scientists interact with matter primarily through machines. They speculate about making machines that will do certain things. Scientists (as well as engineers) construct machines. They test them out, introduce certain kinds of matter, try to get the machines to produce something (lines on a graph, a column turned blue, and so on), and then try to interpret these productions. Much of this process, Pickering argues, is *messy* because matter does not always co-operate. Matter, for Pickering, *resists*. This resistance is the reason that experiments either don't work at all, don't work the way scientists want them to, and/or produce unexpected results. In response, scientists accommodate each resistance by tinkering with their machines, redefining the goal of the experiment, reinterpreting the data, and so on. Central to the mangle of practice is the concept of time: there is a constant back and forth sequence of resistances and accommodations. And while Pickering only seems to refer to matter as resisting, we might also say that scientists resist the ways in which matter speaks, if we think of 'speaking' as the spread of disease, flooding, and so on. This is not to say scientists ever necessarily 'grasp the pure essence of material agency. Instead, material agency emerges via an inherently impure dynamics that couples the material and human realms' (ibid., 54). Pickering refers to this as performativity.

Science, then, is a mangle of matter, machines, and scientists, all responding to each other, all variously resisting and accommodating each other: 'Disciplined human agency and captured material agency are' as Pickering argues, 'constitutively intertwined; they are interactively stabilized' (ibid., 17).

Adrian Franklin (2008) provides an excellent example of how sociologists of science use the concept of the mangle to analyze phenomena. Franklin explores the material exchange between eucalypt gum trees, humans, and fire in Australia. Eucalypt trees constantly shed their branches, and these branches are composed of highly combustible material. Eucalypts are dependent on fire to sexually reproduce because the seed cases of the gum nuts are very tough and have a waxy barrier that melts when exposed to fire. The eucalypt trees, we might say, enjoy an intimate relationship with fire.

Over several hundred years, Aboriginal peoples in Australia learned to practise a system of fire management whereby they actually strategically lit fires in order to manage the path and scope of the fires. Fires were also set to kill other Aboriginal people, concentrate animals that were then killed for food, clothing, and so on. This increased fire changed the composition of plants in the forest as fire-sensitive plants decreased and fire-tolerant plants survived. Then settlers expanded colonial territory, building suburban housing communities within eucalypt forests. Franklin describes the relationship between fire, eucalypt trees, forests (the flora and fauna in these forests and the ecosystem), Aboriginal and Australian peoples, housing, the fire brigades, social policies, and political legislation aimed at better managing farmland and suburban sprawl as a complex mangle in which fire, trees, and ecosystems demonstrate agency as they respond to suburban expansion and varying fire-fighter budgets. The human inhabitants, in turn, respond to the ebbs and flows of fires, eucalypt sexual reproduction, and reforestation.

We have, then, a series of theories that are all concerned to define both what science is (its nature, scope, and underlying assumptions), and how socio-logical theory should be brought to bear on the scientific enterprise. We now turn our attention to two case studies in science that bolster the argument that science is social relations.

Summary

In this chapter we learned sociologists employ a variety of theories in order to better understand why and how science works. We explored Fleck's concept of the thought collective and Kuhn's concept of paradigms. Both concepts draw us away from the traditional view of science as making slow and steady progress towards a particular goal (true and universal knowledge about every-thing), and towards an understanding of science as an indelibly social enter-prise dependent upon particular structures that constrain ways of knowing.

Thought collectives and paradigms are reproduced within each generation of science students in the form of textbooks, lectures, formalized laboratory procedures, course assessment, grant writing, conference presentation, and so on. Thought collectives are also revealed in the myriad informal and tacit ways information about how to do good science is taught to science students. Thought collectives are further revealed in the ways in which scientists specu-late about a given phenomenon (why scientists choose a particular phenom-enon in the first place, and then what particular aspects of that phenomenon they focus on), construct hypotheses, carry out experiments, collect and ana-lyze data, and report their findings. All of these structures and processes com-bine, and by the time students become scientists in their own right, they are fully integrated into a particular paradigm. As such, the bulk of science is what Kuhn refers to as normal science, which consists of working away at problems that conform to a particular paradigm.

Building upon Fleck and Kuhn's analyses of how science works, sociologists propose different theories about how to best analyze science within society. The Strong Programme urges analyses that focus on the discourse—debates, communication, and specialized language that make knowledge possible—of science. Rather than engage with the validity of scientific facts, the Strong Programme focuses on how science establishes knowledge as either credible or discredited. While the Strong Programme has made great strides in high-lighting the social dimensions of science, it has been criticized for ignoring nonhuman agency. In other words, the Strong Programme largely prohibits anything other than humans from exerting any kind of force in how phenomena work. Actor Network Theory, by contrast, forefronts relations between assemblages of human and nonhuman actants. ANT provides a way to take seriously what science takes seriously—objects in the world.

Actor Network Theory has been criticized for its focus on semiotics (language, signs, and communication), which seems to move analyses back into texts and representations—precisely what ANT criticizes the Strong Programme for doing. Andrew Pickering's mangle of practice offers a theory meant to build upon the strengths of SSK to closely analyze the social aspects of scientific practice, as well as draw on the strength of Actor Network Theory to take nonhuman agency seriously without making humans and nonhumans equivalent. The Mangle theorizes science as the interaction between scientists, machines, and objects in real time. This is a performative account of science, 'in which science is regarded [as] a field of powers, capabilities, and performances, situated in machinic captures of material agency' (Pickering, 1995: 7). Pickering refers to machinic captures of material agency because machines, tools, and instruments are essential to scientists' work. As Pickering writes: 'scientists . . . maneuver in a field of material agency . . . constructing machines that . . . capture, seduce, enroll . . . that agency, taming and domesticating it, putting it at our service . . . ' (ibid., 7).

Key Terms

Science Wars   This is the term given to the intellectual debate that took place in the 1990s and 2000s between realists and postmodernists regarding the nature of science. Realists argue that science describes objects in themselves, and postmodernists argue against scientific objectivity.

Scientific realism   This approach to inquiry supposes we may derive factual knowledge about things in themselves.

Plato's Dialogues   *Dialogues* refers to a series of conversations that Greek philosopher Plato wrote to put forth a system of political, economic, and cultural philosophy.

Skepticism   This term refers to a critical attitude towards the truth and/or objectivity of any claim.

Representationalism   This is a philosophical approach to inquiry premised on the supposition that we do not apprehend things in themselves (the world as it is) but rather representations of the world as we perceive and understand it.

Gedanken   This is an experiment that takes place in the mind rather than in the physical world. Gedankens are used by scientists when they do not have the wherewithal to conduct the experiment in real life.

Proletariat   This term loosely refers to people from lower classes who do not own the means of their own productive labour.

Bourgeoisie   This term refers to people from middle classes who own the means of other people's productive labour (such as factories).

Thought collective   This term refers to a community of people who exchange ideas. Thought collectives structure ideas such that they conform to certain parameters and evolve in particular ways.

Paradigm   A thought pattern.

Normal science   Thomas Kuhn used this term to describe the way in which most science operates most of the time within a particular paradigm.

Actant   This term was created by Bruno Latour within Actor Network Theory to describe things (whether humans or leaves) defined through their relations with other actants.

Principle of symmetry   This phrase refers to the principle within the Strong Programme that the same kind of explanation be applied to successes and failures within scientific research. For example, sociologists should evaluate solid-state physics and astrology using the same criteria.

Critical Thinking Questions

1. How do paradigms create matters of fact?

2. How do objects show agency?

3. Does science produce relative or absolute knowledge?

4. How can we judge whether scientific facts are true or false?

Suggested Readings

P.L. Berger and T. Luckmann, *The Social Construction of Reality: A Treatise in the Sociology of Knowledge*. Garden City, NY: Doubleday, 1966. This classic text is frequently used by sociologists because it gives a standard and compelling account of how reality is formed through social practices.

I. Hacking, *Representing and Intervening: Introductory Topics in the Philosophy of Natural Science*. Cambridge: Cambridge University Press, 1983. This book provides an easy-to-read and persuasive account of how scientists construct matters of fact.

T. Kuhn, *The Structure of Scientific Revolutions*. Chicago: The University of Chicago Press, 1962. This is a classic text within science and technology studies. It outlines how normal science operates and under what unusual conditions scientific revolutions take place.

C. Thompson, 'When Elephants Stand for Competing Philosophies of Nature: Amboseli National Park, Kenya', in J. Law and A. Mol, eds., *Complexities*. Duke University Press, 2002, pp. 166–90. This article provides a concrete example of how science and culture work in tandem to create our understanding of what nature is, and what is natural. The book in which it appears is a useful compilation of empirical examples of how science is social.

Websites and Films ··

Science Studies Search Engine
www.google.com/cse/home?cx=006369935143364481409%3Ak8leffjphf8
   This website provides a wealth of sources with diverse publications in the area of
   science and technology studies.

Society for Social Studies of Science
www.4sonline.org
   This website contains information about the 4S, and also relevant publications,
   conferences, and the like.

Science World
http://scienceworld.wolfram.com
   This website provides useful links to publications explaining physics, chemistry,
   mathematics, and astronomy. It also provides useful publications about scientist
   biographies.

How To Think About Science
www.cbc.ca/ideas/episodes/2009/01/02/how-to-think-about-science-part-1---24-listen
   Instead of offering specific suggested films for this chapter, I strongly recommend
   the CBC's podcasts 'How to Think About Science'. You may download all 24
   episodes to your mp3 player, and listen to them on your own. They feature
   very interesting interviews with eminent science studies scholars such as Bruno
   Latour, Lorraine Daston, Margaret Lock, Steven Shapin, and Ian Hacking.

3 | Science is Social Relations: Part II

Learning Objectives

In this chapter we learn:

- ⊕ Physics and biology, as past and present pre-eminent sciences, are strong examples of science as social relations;
- ⊕ The development of quantum theory in physics forced a new understanding of science as a social practice;
- ⊕ Through quantum theory, natural and social scientists revisit key ideas about the subject–object distinction, and the ability of experience and empiricism to access facts about the world;
- ⊕ Modern biology's attempt to reduce life itself to physical and chemical processes is viewed by sociologists within the larger context of science reflecting the desire to control nature;
- ⊕ Research on the cell's cytoplasm and in epigenetics suggests a move towards understanding entities in extended social relations.

Introduction

This chapter continues the argument that science and technology are social relations. Chapter 2 focused on sociological theories that conceptualize science as social practices. The present chapter introduces three fields of research within the natural sciences—quantum theory, genetics, and epigenetics—as further examples of how science and technology are social. Quantum theory represents a Kuhnian paradigm shift away from solid-state physics characterized by the presupposition of the kind Newton advanced that universal laws govern the physical world (see Chapter 2). It precipitates speculation about the validity of the presupposition of scientific laws themselves. Einstein, Heisenberg, Bohr, Schrödinger, and other physicists devoted considerable time and energy to philosophical speculations about the nature of matter, how we represent physical objects, whether facts exist external to human interactions with the physical world, and so on.

Like solid-state physics, modern biology attempted to provide a kind of unified field theory—a theory of everything—that required the reduction of life itself to chemical and physical components that operated via law-like processes. The hegemony of genetics research in the last 40 years reflects the ideal of modern science to understand, and thereby control, nature. This, as the previous chapters showed, is in and of itself, a social relation. And of course, enormous gains in knowledge have accrued within this paradigm. The latter half of this chapter considers some recent challenges to this modern science ideal. Research on the cell's cytoplasm and, more broadly, the nested relationships of genes to cells, cells to organism, organism to communities of organisms, and communities to dynamic environments suggest a more holistic view of life. The claim here is not optimistic: science, as Chapter 5 will show, remains largely co-extensive with the political economy of capitalist western society.

Quantum Weirdness

Physics has a long tradition of trying to ascertain solid and timeless laws governing the universe. This is sometimes referred to as classical physics, Newtonian physics, solid-state physics, or the physics of being.

Plato and Aristotle, as philosophers of nature as well as society, were interested in the world as a solid and ageless being. Plato understood the world as made up of timeless, unchanging external objects whose essence could be understood by comprehending that object's ideal standard (Dupré, 1996; Ghilselin, 1997). An object's essence, in other words, is the set of attributes said to make up an object. Plato argued the objects we encounter are actually imperfect *copies* of themselves; objects themselves have timeless and unchanging essences. So for Plato, studying any given object could not, in itself, explain that object. To explain an object's essence, we must, according to Plato, turn away from nature, and toward philosophy. Only philosophy could derive an object's true inner being or essence.

Aristotle, in turn, distinguished philosophy from physics. He argued physics is concerned with things that exist in nature, forces such as water and air, as well as plants and animals (Waterlow, 1988). Aristotle argued natural things are self-contained in the sense that stasis and change are contained within the object of nature itself, and do not rely on forces outside nature (Waterlow, 1988; Lang, 1992). Natural things that have been transformed by humans— metal into a frying pan for instance—become things outside nature. These things outside nature are the purview of philosophy.

Fast-forward to the seventeenth century and we find Isaac Newton (amongst other naturalists, philosophers, and explorers) hard at work devising a quantitative (mathematical) explanation for what Aristotle and Plato had only qualitatively described. For example, Newton wanted to devise a mathematical formula that would allow him to predict the motion of objects in a system (Auletta, 2006). In order to predict motion or other forces, Newton needed to control the system, which he did by defining a closed system as that for which every factor can be taken into account by the observer. Such a closed system could produce quantifiable and predictable cause–effect relationships about a given force acting on an object and that object's change in motion (Cohen, 1980). For Newton, matter is substance 'in its own right . . . as being' (Houghtaling, forthcoming). In other words, in classical physics objects have a constant identity; an observer can measure 'this particular property of this particular object at any given point or state' (Houghtaling, forthcoming; Heylighen, 1990).

Experiments in classical physics work with an idealized version of nature in the sense that nature is temporarily and artificially closed and controlled. Experiments say 'if this, this, this, and this are controlled, the object will do this.' As such, classical physics necessarily reduces phenomena. We can see the appeal: classical physics provides us with a way to understand things like gravity, motion, light, sound, energy, mass, and so on. Being able to understand how things work and predict what will happen in our environment increases our confidence that we are that much closer to the truth and that we will not be relentlessly surprised in our interactions with the environment. And it is no coincidence that the foundation of sociology as a discipline emerged from classical science in the form of positivism (see Chapter 2). Hume (1739/2000), Mach (1887/2009), Comte (1853/2009), and Durkheim (1895/1982), amongst others, embraced the scientific method as the best means of studying the causes of social phenomena, from suicide to marriage to government. Positivism sought to understand the *causes* of societal change and transformation.

In the 1930s a series of discussions began to take place amongst physicists. These discussions revolved around the genesis of quantum physics, and the discussants included luminaries such as Albert Einstein, Niels Bohr, Werner Heisenberg, and Erwin Schrödinger. The discussions certainly involved a lot of complicated math. But they also involved serious discussions about the philosophy of science, including what is known, what can be known, and how we know things. These discussions included what are now known as the Uncertainty Principle, the Copenhagen Interpretation, Schrödinger's Cat, Occam's Razor, and so on.

The basis of the discussion was how to reconcile what happens at the normal scale of things—meaning things at a scale that our human eyes can perceive such as lakes, grass, and raindrops—and the quantum level, or really small things like atoms (that humans cannot see with the unaided eye). Physics maintains there is a single world, and as such, there cannot be two (quantum and macro) conflicting realities (this is called the Correspondence Principle).

Things started to get weird when physicists contemplated something called the Wave-Particle Duality Paradox. Particles are very small bits of matter. Each bit of matter occupies a particular point in space and time. Waves, on the other hand, are not things per se, but more like disturbances (which cannot be localized to a point) that propagate in a medium (like water) or as oscillating fields (like electromagnetic waves such as light). For instance, think of the

BOX 3.1 ⁕ GEDANKEN

A *gedanken*—derived from the German word for thought—refers to a thought experiment. They are used in science to both generate and test hypotheses. Gedankens are a useful tool for thinking about experiments that cannot be performed for a variety of reasons (for example, if it is not technologically possible or if it would be unethical to perform the experiment). Famous gedankens include Maxwell's Demon and Schrödinger's Cat.

In the 1870s, James Maxwell derived a gedanken to argue the second law of thermodynamics was not actually a law but rather a statistical probability. The second law of thermodynamics states that two entities brought together in an isolated system will develop equilibrium. Think about what happens when you run a bath. To get the temperature right, you add both hot and cold water. The hot and cold water combine to evolve equilibrium. Maxwell derived a gedanken in which there is a sealed box (an isolated system) containing two rooms separated by a wall with a door that can be opened and closed. Controlling the door is what Lord Kelvin called Maxwell's Demon—an entity that has complete control over the door and what is able to pass through it. By letting a single molecule at a time through the door, the demon would be able to create a situation in which the side that the molecule has just left will be slightly cooler than the side that the molecule has just entered (since speed corresponds to temperature). Numerous physicists have since critiqued this gedanken.

Erwin Schrödinger produced the famous gedanken known colloquially as Schrödinger's Cat. The gedanken was meant to point out the logical problem with quantum theory that outcomes exist as probabilities rather than certainties. A cat is placed in a sealed box containing a table on which rests a Geiger counter and a radioactive source. If the Geiger counter detects decay from the radioactive source (perhaps a single atom decays), it releases poison, which kills the cat. If the machine does not detect decay, it does not release poison, and the cat remains alive. The problem is that we cannot know whether the cat is alive or dead until we open the box and see. Since opening the box and looking at the cat requires the interaction of the box–cat-observer system, until the box is opened the cat is simultaneously alive and dead. The fate of the cat is 'entangled' with the fate of the atom. Schrödinger's point is that a simultaneously alive and dead cat defies common sense, which holds that the cat is *either* alive *or* dead but not both simultaneously.

waves created by dropping a stone in a lake. Waves, under the right conditions, produce diffraction patterns when they combine, overlap, and/or encounter an obstruction (Barad, 2007).

Given these definitions, and thinking from a Newtonian perspective, physics would predict that only waves would produce diffraction patterns because particles cannot occupy the same place at the same time. To think this through, Niels Bohr adapted Thomas Young's two-slit experiment to produce a *gedanken*, or thought experiment, to see what happens at the quantum (really small) scale of things. In this gedanken, a machine shoots particles through two equally sized openings (slits) in a partition (such as a wall), and the particles that go through the openings then hit a wall on the other side. If particles behave in the way they do at the normal macro level (at a scale that we can see them with the naked eye), shooting them through the openings produces the pattern shown in Figure 3.1 below.

As the diagram shows, the particles either hit the partition or they pass through one of the openings and hit the wall to form a line. Shooting particles produces a pattern we expect (the diagram on the left), and waves produce a pattern we expect (the diagram on the right). Moreover, the same pattern occurs when we shoot particles through two, three, or more slits, and regardless of whether or not we shoot the particles one at a time or in bunches.

At the quantum level, *when the experiment is observed*, shooting particles through the slits produces a particle distribution (as seen in the diagram on the left). *When the experiment is not observed*, particles behave like waves. That is, shooting particles (and it doesn't matter whether we propel them as a bunch or one at a time) through one or more slits produces a diffraction pattern like the diagram in Figure 3.2.

In classical Newtonian physics electrons, atoms are *either* waves or particles, independent of their observation or other experimental circumstances. So how can a single electron interfere with itself to produce a wave pattern? Even more strangely, when defined mathematically, it turns out that a single electron can go through no slit, one slit, or both slits at the same time

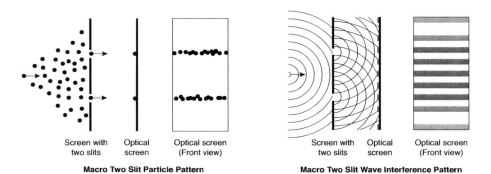

Screen with   Optical   Optical screen   Screen with   Optical   Optical screen
two slits   screen   (Front view)   two slits   screen   (Front view)

Macro Two Slit Particle Pattern   **Macro Two Slit Wave Interference Pattern**

FIGURE 3.1 The Two-Slit Experiment: Results Expected Within Classical Physics Paradigm
(Image courtesy of Anthony Krivan)

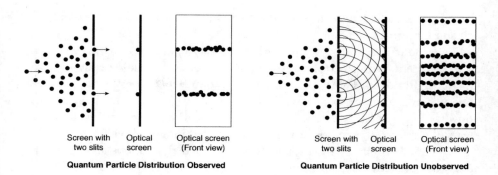

FIGURE 3.2 Two-Slit Experiment: Quantum Level
(Image courtesy of Anthony Krivan)

(Barad, 2007: 83). How is this possible? And why does it matter whether the experiment is observed? Keep in mind that the observer doesn't need to be human. Also keep in mind that this result has been observed with electrons, neutrons, atoms, and other quantum matter, so it doesn't matter what kind of matter we propel through the openings.

Here's where physics meets philosophy. The more conservative interpretation is an explanation about how we know what we know. Werner Heisenberg came up with this interpretation, known as the Heisenberg Uncertainty Principle. For Heisenberg, the particle-wave duality paradox empirically demonstrates that we can only make probabilistic predictions about energy/time and position/momentum. That is, before we shoot the particles through the slits, we can only say that they will probably behave in a particular way. It is only when we have seen the results of the experiment that we can say that the particles did behave in particular ways. We can't be certain, according to Heisenberg, before or during the experiment—only after the experiment.

Thus, for Heisenberg, Einstein, Podolsky, Rosen, and others, actually observing something (particles moving through a two-slit experiment, or the moon, or anything else) produces definite results: before something is observed, it has a range of probabilities. As Heisenberg wrote 'we do not have a science of nature, we have a science of our *description* of nature' (in Thompson, 1991: 14, emphasis added). To the question 'does the moon exist apart from human observation?' the answer, according to this interpretation, is that we can only make meaningful utterances when someone has actually looked at the moon. For Heisenberg, it is a measuring problem.

As if this explanation wasn't already complicated, Bohr thought the situation was even more complex. He developed the Copenhagen Interpretation to say that the wave-particle duality paradox means that particles do not have a simultaneously determinate position *and* momentum value. That is, wave and particle behaviours are *complementary*—exhibited in mutually exclusive circumstances.

What Bohr is arguing, in opposition to Heisenberg, is that this is not a measurement problem to be resolved with greater technologically precise instrumentation or more objective observation procedures (i.e., it is not about doing a better experiment). For Bohr, this is not a problem about how we know what we know: it is a problem of what we know. Put another way, it is the difference between uncertainty and indeterminacy.

In 1935, Einstein, Podolsky, and Rosen published a famous article which distilled Einstein's concern about what counts as reality. In the following quote, Einstein describes what he holds to be reality: that if two objects (A and B) are in two different spaces and times, then somehow changing A should not affect B. Einstein was so concerned about the implications that he articulated quantum theory as a crisis in defining what the discipline of physics is actually all about:

> I just want to explain what I mean when I say that we should try to hold onto physical reality . . . That which we conceive as existing ('actual') should somehow be localized in time and space. That is, the real in one part of space, A, should (in theory) somehow 'exist' independently of that which is thought of as real in another part of space, B . . . What is actually present in B should thus not depend upon the type of measurement carried out in the part of space, A; it should also be independent of whether or not, after all, a measurement is made in A . . . If one renounces the assumption that what is present in different parts of space has an independent, real existence, then I do not at all see what physics is supposed to describe. For what is thought to be a 'system' is, after all, just conventional, and I do not see how one is supposed to divide up the world objectively so that one can make statements about its parts. (1935: 696)

The crux of Einstein, Podolsky, and Rosen's argument is that if we can control a system such that it is either not disturbed at all or that its total disturbance can be measured and accounted for, we can retain a notion of causality, and thus reality, as defined by classical physics. Bohr responded that in classical physics we may talk about causation, but that in 'quantum theory the uncontrollable interaction between the objects and the measuring instruments forces us to a renunciation even in such respect' (1961: 60–1). In other words, in quantum theory, we may only talk about probabilities.

Bohr argues we cannot separate objects from their observers (human or otherwise). To take the experimenter out of the experiment is to accord the experimenter with a privileged position (recall Pickering's mangle of practice that includes humans; see Chapter 2). So Bohr defines what Einstein calls reality as *phenomena*, defined as 'the observations obtained under specified circumstances, including an account of the whole experimental arrangement' (1961: 50). In other words, reality is defined as things-in-phenomena rather than things-in-themselves (von Weizsacker, 1980).

In *Meeting the Universe Halfway* (2007), physicist turned philosopher Karen Barad uses the term intra-action to refer to the inseparability of all words (culture) and all things (nature), contrasted against the term inter-action and predicated upon individuated entities that subsequently interact. Barad defines realism as 'not about representations of an independent reality but about the real consequences, interventions, creative possibilities, and responsibilities of intra-acting within and as part of our world' (2007: 37).

Barad's sophisticated book is an extended conversation about what we know and how we can know things. One important implication is our understanding of objects. For Barad, 'the primary . . . unit is not independent objects with inherent boundaries and properties but rather phenomena' (2007: 139). 'Existence', in other words, 'is not an individual affair . . . individuals do not pre-exist their interactions; rather, individuals emerge through and as part of their entangled intra-relating' (2007: ix). This incessant intra-action takes place with or without a conscious observer.

There is no shortage of speculations about what all this means. Just Google 'quantum weirdness' or 'quantum theory' to get a cascade of sites devoted to arguing that quantum theory proves the existence of multiple universes, that there is no reality, and so on. Or watch the 2004 film *What the Bleep?* Ostensibly drawing from quantum theory, this film argues people are entirely responsible for what happens to them. Poverty, violence, child sexual abuse, and obesity are all, according to this film, a product of individuals thinking these things into reality (because everything is connected, the argument goes, humans can influence everything that goes on around them). So, people are poor because they think themselves poor, according to the film.

Barad (like Bohr) is careful to distance her position from those who interpret quantum theory as advocating a kind of everything-is-connected-so-therefore-everything-is-within-our-power-to-change position. She writes, '. . . one can find suggestions in the literature that quantum physics is inherently less androcentric, less Eurocentric, more feminine, more postmodern, and generally less regressive than the masculinist and imperializing tendencies found in Newtonian physics' (ibid., 67). However, as Barad reminds us, quantum theory led to the development of the A-bomb, particle physics is scientific reductionism at its best in some respects, and quantum theory generally remains the purview of a handful of primarily western-trained males. Moreover, it is a testament to human exceptionalism that we would place ourselves at the centre of the universe and assume that we are the ones forever in control.

For sociologists, quantum theory and the discussions that ensued by physicists and philosophers, lead to a number of implications. First of all, 'physical laws refer not to "nature" but to our relation to it. Facts, then, are theory, language, and technique laden, making *relations*, not things, the true object of inquiry in contemporary science' (Aronowitz, 1988: 250). Science, in other words, is social relations. 'Interaction', as Aronowitz states, 'is the condition of [science's] construction' (ibid., 248).

Second, it means that facts do not simply and unproblematically exist. Matters of fact, as Chapter 2 argued, are created. What becomes a fact is the outcome of a complex intra-action of science as a set of presuppositions (including measurement, apparatus, observer, experiment, etc.). As such, 'science is a self-referential inquiry' (ibid., 244). That is,

> We do not merely 'observe' phenomena, but phenomena are constituted, in part, by the consequences of intervention. The subject/object split becomes as untenable as the materialist hypothesis that our knowledge refers to an external world that is self-contained. (ibid., 246)

Third, it means that the distinction between the natural and social sciences is unsustainable because both sciences are subject to self-referential conventions of language and norms. Charles Sanders Peirce goes so far as to argue truth is what a 'legitimate' scientific community agrees it is (1965: 65).

Life Itself

A commonly held view is that scientists are interested in proving that nature trumps nurture in all matters of life. This reductionism finds voice in sound bites such as 'it's all in our genes', the human brain is no more than a neuro-transmitting device, and that bodies may be explained *in toto* by molecular and physical determinants. Back in 1866, Gregor Mendel established that genes played a part in the inheritance of particular characteristics in pea plants. It was not, however, until Rosalind Franklin, Maurice Wilkins, James Watson, and Francis Crick produced the now-famous double-helix structure of genes in the late 1940s and early 1950s that scientific work and public interest in these minute chemical chains really took off (see Chapter 5, Box 5.1). Evolutionary theorists quickly embraced genetics within what is known as the modern synthesis (Darwin's evolutionary theory combined with genetics), creating a better understanding of how species evolve through natural selection (Huxley and Baker, 1974). Today, many people understand genetics research to substantiate a sort of 'genes-r-us' formula.

This formula largely derives from what Crick (1958) termed the Central Dogma. The Central Dogma states that DNA makes RNA; RNA makes proteins; and proteins make us.

John Protevi (2008) summarizes the Central Dogma as a step-wise process:

BOX 3.2 ❄ WHAT IS DNA? WHAT IS RNA?

DNA stands for deoxyribonucleic acid. DNA stores information that plays a role in the development and functioning of all living organisms and some viruses. DNA segments are called *genes*. The well-known image of the double helix is composed of two long polymers called nucleotides connected by bonds made of sugars and phosphates. Each sugar is attached to one of four bases: adenine (A), cytosine (C), guanine (G), and thymine (T). Adenine binds with guanine and cytosine binds with thymine.

RNA stands for ribonucleic acid. It is a molecule made up of a long chain of nucleotides composed of a nitrogenous base, a sugar, and a phosphate. RNA and DNA are very similar. However, they differ in that RNA is typically single-stranded, RNA contains ribose and DNA contains deoxyribose (it lacks an oxygen atom), and RNA has uracil rather than thymine.

A genome refers to all the genes in an organism. Genotype refers to a cell's genetic composition. A phenotype refers to an organism's observable characteristics (such as eye colour).

---

1. In the nucleus, two strands of DNA pull apart.
2. Through a process known as *transcription*, an RNA polymerase enzyme copies one of the DNA strands into a strand of mRNA.
3. The mRNA strand is transported out of the nucleus and into the cytoplasm.
4. Through a process known as *translation*, transfer RNA (tRNA) binds to the mRNA.
5. Through a process known as translation, the tRNA adds an amino acid monomer to the polymer chain.
6. The protein chain then moves into the cell to direct activity there.

For our purposes here, the important point about the Central Dogma is that it characterizes cell development and function as a one-way process: the flow of information goes from DNA to proteins to the organism as whole. In this formulation, scientists conceptualize DNA as master molecules that direct, and are the blueprint for, life. This means that evolutionary variation (how we get changes in organisms over time) is only produced through random genetic mutation.

Watson, Crick, and Wilkins's 1953 Nobel Prize–winning research effected a powerful unification in the biological sciences: just as atoms became the common language of physics and molecules became the common language of chemistry, DNA became the signature of biology, leading quickly to sub-fields of research (molecular genetics is the most obvious one) and well-publicized research such as the Human Genome Project (HGP) (see Chapter 5).

The HGP began in 1990 and was finally completed in 2003 by the UCSC Genome Bioinformatics Group. When it was announced that scientists had mapped the genome, it came with several caveats, not the least of which was disagreement as to whether the entire sequence had really been identified. It essentially led to more scientific questions than answers about the complex ways in which genes interact with their environment; that is, the cytoplasm, the body, and the body's environment (Lewontin, 2000). There is an emerging appreciation for the complexities and nuances of molecular processes. Scientists themselves are challenging the Central Dogma through ongoing research.

We are beginning to appreciate that genes are heavily regulated by, and indeed controlled by, cytoplasmic material (all the other stuff in the cell such as proteins and other enzymes) (Keller, 2000, 2002). Consider eyesight, for example. When a mouse's gene for eyes is transplanted into an eyeless fly, it induces eye formation. We get a mouse gene that works in a fly. The eye that forms, however, is a fly eye. So it's what we might call the fly context—the cytoplasmic proteins, enzymes, and so on—that determines the *kind* of eye that is formed (mouse or fly) (Halder, Callaerts, and Gehring, 1995).

Increasing evidence suggests the importance of proteins, the cytoplasm, and the wider environment in determining which genes are activated, transported, copied, and spliced; indeed, these developments are leading geneticists to coin a new central dogma, 'one protein—many functions'. As Jablonka and Lamb write, 'the gene cannot be seen as an autonomous unit . . . a "gene" has meaning only within the system as a whole' (2005: 7). Indeed:

> left to its own devices, DNA cannot even copy itself: DNA replication will simply not proceed in the absence of the enzymes required to carry out the process. Moreover, DNA is not intrinsically stable: its integrity is maintained by a panoply of proteins involved in forestalling or repairing copying mistakes, spontaneous breakage, and other kinds of damage incurred in the process of replication. Without this elaborate system of monitoring, proofreading, and repair, replication might proceed, but it would proceed sloppily. Accumulating far too many errors to be consistent with the observed stability of hereditary phenomena—current estimates are that one out of every hundred bases would be copied erroneously. With the help of this repair system, however, the frequency of mistakes is reduced to roughly one in 10 billion. (Keller, 2000: 26–7)

The stability of gene structures depends upon the complicated orchestration of the cell as a whole (Keller, 2000). Cells employ a complex range of mechanisms that scientists assign equally complicated terms to: alternative splicing, crossing over, posttranscriptional editing of mRNA, repair mechanisms that sacrifice fidelity, enzyme-mediated modification of protein composition, chromatin diminution, polyploidy, polyteny, amplification, rearrangement, homeotic genes, lateral gene transfer, enzyme reverse transcriptase, and other phenomena (Gilbert, 2002; Jablonka and Lamb, 2005; Waddington, 1953).

To give flavour to what I mean by 'other phenomena' consider *nucleotide replacement*. In a sequence of only 100 units (and DNA strands are often much longer), made up of four modules (adenine, thymine, cytosine, and guanine), 4^{100} different sequences are possible. This is more than the number of atoms in the galaxy. Sickle cell anemia is caused by a single nucleotide change (an A instead of a T) in the nucleotide sequence in hemoglobin (Jablonka and Lamb, 2005: 55). In other words, the difference between not having and having sickle cell anemia is:

CTG	ACT	CCT	GAG	GAG	AAG	TCT
CTG	ACT	CCT	GTG	GAG	AAG	TCT

BOX 3.3 ❖ TRANSPOSONS AND THE SOCIOLOGY OF SCIENCE

Barbara McClintock, an American cytogeneticist who studied maize (corn), discovered transposons. Through very careful study over many years, McClintock made a number of path-breaking discoveries. She was able to see chromosome structures and sequencing through microscope techniques she developed herself. McClintock was the first person to provide a genetic map for maize and to explain how telomeres and centromeres (chromosome regions) work in the cell to preserve genetic fidelity. She was able to demonstrate genetic recombination by crossing-over during meiosis. Perhaps her most important research was to show how transposons work. Once called jumping genes, transposons are sequences of DNA that move around within the genome of the cell, changing the amount of DNA in a cell and causing mutations.

McClintock earned her PhD in the 1920s and performed her research in the 1940s and 1950s within a male-dominated discipline and laboratory settings. Her research, like that of other female scientists, was not taken seriously for many years to such an extent that McClintock even stopped publishing her research in the 1950s. She was elected to the National Academy of Sciences in 1944, and earned the Nobel Prize for Physiology or Medicine in 1983 for her work on transposons. McClintock is the only female scientist to have been awarded an unshared Nobel Prize in Physiology or Medicine. Evelyn Fox Keller, a Science studies scholar, wrote a richly detailed book about McClintock's research in which she argues that McClintock's identification and treatment by male colleagues through the years as a specifically female scientist had an impact on her research (see *A Feeling for the Organism* [1983]).

What this list of mediating factors shows is that cellular activity is a really complicated system: it involves proteins, enzymes, genes, and the cell as a whole, as well as the cell's dynamic interaction with its environment. In other words, scientists derive limited purchase from reductionism and the ideal of science to provide a unified field theory—a kind of theory of everything—of life itself. Instead, research in biology suggests the merits of going bigger; of thinking of genes within cells, cells within organisms, organisms within communities of other organisms, and communities within changing environmental contexts. The recent turn within molecular biology to the study of epigenetics may be a sign that the hegemony of reductionist biology has some challengers.

The modern synthesis focuses on how changes in an organism's genotype (characteristics produced by genes) produce changes in that organism's phenotype (physical characteristics we see such as hair colour). The modern synthesis maintains that such changes take place *only* through random mutation. Epigenetics challenges this theory.

Epigenetics refers to genetic changes in organisms over generations through non-DNA means. It is how organisms express different traits without changes to their DNA. While there is no change in the underlying DNA sequences from this process alone (changes occur through other means as we saw above), non-genetic factors determine gene expression. As such, epigenetics challenges the Central Dogma.

Epigenetics is gaining increasing attention within evolutionary and development biology, or Evo-Devo as it is known (Protevi, 2008). Researchers have known for some time that identical genes can lead to very different phenotypes, and that disabling genes known to influence particular developmental pathways can make no difference to the final phenotype (Jablonka and Lamb, 2005). Now we know that most specialized cells are epigenetic, and acquire information that they pass to their daughter cells through epigenetic inheritance systems (EISs).

Epigenetics was discovered by a group of scientists investigating the inherited effects on children born to women who endured periods of starvation during pregnancy. Bastiaan Heijmans and colleagues analyzed detailed records kept by a Dutch village during the 1944 winter when Germans occupied the Netherlands and cut off supplies to their village. All people during this time were rationed to about 1,000 calories per day, dropping to about 500 calories per day towards the end of the blockade. The circumstances were somewhat unique because the village's registry kept detailed records of births, deaths, and food intake, and the famine took place over a specific documented period of time. The researchers divided the records into three populations: (1) people who were conceived or born during the famine, (2) same-sex siblings who were born before or conceived after the famine, and (3) unrelated people (control group). The scientists measured the IGF2 gene—which is involved in human growth and development. More methylation means less IGF2.

Sixty years after the famine, people who were conceived during the famine had less methlyation on their IGF2 gene compared with their unexposed same-sex sister or brother. These people also had worse glucose tolerance, higher blood pressure, and higher rates of obesity and hypercholesterolaemia. Scientists think this increased cardiovascular and metabolic risk is caused by the body's increased efficiency in storing calories. When the famine ended, these people (as either fetuses or children) experienced a dramatic increase in their daily caloric intake. Their bodies could not stop storing the calories, which caused cardiovascular problems (see also Lumey, 2007; Roseboom et al., 2000a, 2000b).

What this study showed was that changes in the environment effect people's genes in terms of what genes are activated and deactivated. Swedish scientists have found that people have higher mortality risk ratios if their paternal grandmother or grandfather starved as teenagers, but only if they are the same sex as the grandparent who starved (Pembrey et al., 2006). Another study found that children born to women post–gastric bypass surgery (surgery that makes the stomach smaller) are 52 per cent less likely to be obese compared to their brothers and sisters born while their mother was obese. So surgery, and not a change in lifestyle, affects obesity in children (Kral et al., 2006).

Epigenetics has been able to show that genetic variation is not entirely random. In other words, epigenetics suggests that what is known as soft inheritance from the environment is *also* at play in natural selection (Jablonka and Lamb, 2005: 7). Epigenetics joins an increasing list of phenomena that is providing nuance to the Central Dogma, and therefore to our understanding of evolutionary change in general.

Summary

Chapters 2 and 3 demonstrate that science is social. Chapter 2 concentrated on theories developed by sociologists that provide a framework for exploring science as a complex set of social practices. Actor Network Theory and the mangle of practice are, amongst others, sociological theories that call for the need to acknowledge nonhuman entities as constituent elements of what makes science social. To paraphrase Latour, all science is social, but not all sociality is human. Chapter 3 provides an important expansion of this argument by exploring two fields within the natural sciences that emphasize social relations.

Physicists' interest in quantum physics is not only a change in the popularity of fields of study within physics, but a paradigm shift. The movement within physics away from solid-state to quantum theory research is, as Fleck would call it, a change in thought collectives. Solid-state physics was made possible within a world view within which the universe conforms to universal and unchanging laws. Science, within this world view, is tasked with discerning these laws, primarily accomplished by controlled system experiments. But, as Aronowitz observes, 'We are witnessing a slow, discontinuous breakup of the old world-view according to which physical science offers context-free knowledge of the external world, knowledge whose certainty may be posited as a cultural ideal to which other disciplines should strive' (1988: 265).

Science as social relations becomes transparent through quantum theory because the observer is entangled with what is being observed. Thus, *science explores the social relations between entities.* In this sense, quantum theory satisfies sociological theories such as Actor Network Theory and the mangle of practice by studying the social relations between human and nonhuman entities. We are left with a number of questions such as whether or not it is possible to study anything that does not involve humans (because to study something means we are always at least observers and therefore entangled in the phenomenon we are studying).

The development of modern biology to some degree parallels the shift from solid-state to quantum theory within physics. Modern biology is characterized by the ascendancy of molecular biology—genetics research—and its characterization of DNA as the blueprint for life itself. Determining and defining this code as the baseline for all life reduces life to a set (however complex) of discernible law-like rules and processes. As such, genetics research and solid-state physics share the same ideal of modern science (see Chapter 1). Molecular biology's partial shift towards research on cell cytoplasm as well as understanding genes within the context of cells, organisms, communities of organisms, and shifting environments acknowledges and emphasizes the interactions (entanglements) between entities. Epigenetics provides an excellent example of a research program that explicitly studies the intra-actions between genes, organisms, and their environments. This does not mean, necessarily, that molecular biologists are engaged in philosophical arguments about the entanglement of phenomena, but it does suggest a new emphasis on social relations as the object of biological research.

Key Terms

Theory of everything   This term originated in physics and refers to the creation of a theory that would be capable of describing every physical process in the universe and predicting every physical experiment. So far, physicists have been unable to derive a generally accepted theory of everything.

Complementarity   Derived by Niels Bohr in his discussions of quantum theory, this term refers to the idea that the behaviour of a system cannot be separated from the instruments used to measure that system.

Indeterminacy   This is a term used in quantum physics to refer to the idea that descriptions of physical systems are necessarily incomplete because objects at the quantum level have indeterminate states.

Central dogma   This refers to the basic tenet within early molecular biology, which states that once information has been transferred from DNA to RNA and then to proteins that it cannot move back to DNA.

Translation   This term refers to the third stage in the biosynthesis process whereby information contained in RNA is decoded by ribosomes into amino acids (proteins).

Transcription   This term refers to the second stage in the biosynthesis process whereby a DNA sequence is copied into an RNA sequence.

Human Genome Project (HGP)   The Human Genome Project is an international consortium of scientists attempting to decode the entire complement of gene sequences in the human being.

Modern synthesis   Sometimes referred to as the evolutionary synthesis or neo-Darwinian synthesis, this term refers to the union of evolutionary theory with genetics.

Epigenetics   This term refers to the study of *inherited* phenotypic changes in organisms caused by mechanisms other than changes in the DNA sequence.

Critical Thinking Questions

1. If a sociologist and a physicist were to work together on the problem of 'entanglement', what might their research question(s) be?

2. If a sociologist and a molecular biologist were to work together on a problem concerning the determinants of human heath, what might their research question(s) be?

3. How might sociologists help scientists to better understand the ways in which publics take up and use scientific research related to quantum theory and genetics?

4. Is it possible to research the interactions between entities without the interaction of humans?

Suggested Readings

E. Fox Keller, *The Century of the Gene*. Cambridge, MA: Harvard University Press, 2000. This book provides an excellent critical social scientific analysis of the gene as an object of scientific study, metaphor, and symbol within western societies.

T. Hey and P. Walters, *The New Quantum Universe*, 2nd edn. Cambridge, MA: Cambridge University Press, 2003. This book provides an accessible introduction to quantum theory within physics and its implications for understanding the universe.

E. Jablonka and M. Lamb, *Evolution in Four Dimensions: Genetic, Epigenetic, Behavioral, and Symbolic Variation in the History of Life*. Cambridge, MA: The MIT

Press, 2005. This interesting and compelling book argues the evolution of life is best understood as the interaction of myriad genetic, environmental, behavioural, and symbolic factors.

M. Laubichler and J. Maienschein, *From Embryology to Evo-Devo: A History of Developmental Evolution.* Cambridge, MA: The MIT Press, 2007. This book provides an accessible introduction to biological research focused on the interaction between evolution and development.

Websites and Films

John Protevi's Web Site
www.protevi.com/john
John Protevi's three lectures on Deleuze and Biology provide an excellent introduction to 'evo-devo'.

PBS: What is Evo-Devo?
www.pbs.org/wgbh/nova/beta/evolution/what-evo-devo.html
This PBS NOVA website provides a wealth of information about 'evo-devo'.

The Elegant Universe
This three-part DVD produced by PBS NOVA focuses on the search for Einstein's dream: a unified theory of physics.

4 How Science is Social

Learning Objectives

In this chapter we learn:
- All research, whether scientific or sociological, involves theory and methods working in tandem;
- Sociologists study how science works in practice, emphasizing science as a social practice;
- Sociologists study science and technology using a number of methods, including ethnography and discourse analysis.

Introduction

Sociologists studying how scientists know what they know begin from the premise that science and technology are social practices. Chapter 1 explored sociologists' claim that science and technology have never been modern or pure. By this sociologists mean that the goals of modernity to sever humans from everything else in the universe and to produce a set of universal, certain, and true theories and methods to both study the universe and accumulate knowledge about it have never been (nor could they be) reached. Chapter 2 went on to describe the major contemporary ways in which sociologists theorize science and technology. Whether from a Strong Programme or Actor Network Theory perspective, sociologists understand science and technology to be fundamentally social practices. Where sociological theories differ is whether or not, and to what degree, nonhuman entities are included in this social practice.

This chapter focuses on *how* sociologists analyze the sociality of scientific practice. Sociologists employ a number of methods—archival research, discourse analysis, ethnography, observation, participant observation, interviewing, etc.—to explore how scientists practise science. In some instances, sociologists study scientists and produce critical biographies illustrating the social context in which the particular scientist operates or operated. Evelyn Fox Keller, for instance, provides a superb analysis of Nobel Prize–winning plant geneticist Barbara McClintock's research practices within a male-centred science milieu.

Other sociologists focus on how scientists work in laboratories. Laboratories consist of a number of interrelated components, and are a key site of scientific practice. It is in laboratories that scientists make up what Shapin and others call matters of fact (see Chapter 2). Sociologists such as Ian Hacking demonstrate that science depends upon practices such as the construction of models (simplified versions of reality whereby scientists try to control the variables in a given environment), learning to see phenomena, transcribing phenomena into data, and inscribing data with meaning. Scientists and engineers also devote considerable time and energy to tinkering with machines and other instruments, data, hypotheses, and so on. All of these social practices depend upon tacit knowledge, which itself is an inherently social phenomenon.

In order to explore how sociologists study science and technology as social practices, we must first come to grips with the overarching method that scientists use to study phenomena: the scientific method.

The Scientific Method

Since the entire scientific enterprise, from astrophysics to genetic tissue engineering, is based on the scientific method it is important to begin our analysis by considering the explicit rules modern scientists adhere to when they conduct research. The scientific method is a set of techniques for studying phenomena. These techniques attempt to ensure reliability and validity. A system is reliable if it performs in the same way over and over again. Scientists routinely repeat experiments to judge how reliable the knowledge derived from that experiment is. If a scientist repeats an experiment over and over again and gets the same results, it indicates the experiment is reliable. If a different result is obtained each time, this indicates the experiment is unreliable. Validity means that a given idea will produce knowledge that logically follows from this idea. Scientists are always concerned to produce knowledge that is valid—knowledge that follows logically from the premise of a given experiment or theoretical proposition.

The end goal of the scientific method is always to acquire new knowledge about phenomena. Scientists derive this new knowledge by gathering data. Data are groups of information. In order for scientists to use data, they must be both observable (scientists need to be able to detect data) and measurable (scientists must be able to take account of the information). The concept of the soul does not constitute scientific data because the soul cannot be observed and it cannot be measured. Data must also be subject to reasoning. Information that relies entirely upon something else—revelation for instance—does not count as data within science.

The scientific method follows this sequence:

1. Define a question. A research question must be something that can actually be studied.

2. Observe and measure the phenomenon. Scientists must be able to collect information about the phenomenon they want to know about.
3. Form a hypothesis. Scientists must be able to form hypothetical explanations of the phenomenon they are studying.
4. Perform experiment and collect data. Scientists must be able to test their hypotheses by performing experiments. Experiments manipulate objects in particular ways in order to discern how the phenomenon under study responds.
5. Analyze data. Scientists scrutinize the data collected from experiments in order to judge to what degree the results of the experiment correspond to the predictions made by their hypothesis.
6. Interpret data and draw conclusions. Scientists interpret the results of their experiments in the context of any other research undertaken (by themselves as well as other scientists) on a given phenomenon. Scientists then draw conclusions about the validity and reliability of their study, and what new knowledge may or may not have been derived from their experiment.
7. Publish results. A crucial stage in any research program is the publication of research. Publishing results ensures other scientists gain access to the procedures scientists followed in performing experiments. Publishing puts data and its interpretation into the public domain, thereby enabling peer review and public scrutiny. Publishing is, therefore, a way that scientists increase the reliability and validity of their research.
8. Retest. A scientific experiment is rarely carried out only once. Typically, scientists refine their hypotheses based on the results of previous studies (their own and other scientists), and then re-test either using the same variables or altering them.

The historical development of the scientific method reveals an interesting relationship between theory and method. Ian Hacking (1983), Canada's pre-eminent philosopher of science, draws attention to a long-standing prejudice for knowledge derived from representing phenomena (theory) rather than intervening with phenomena (method). Part of this prejudice, according to Hacking, produced the idea that theory always comes before method: that scientists hypothesize something before they test it out. (This prejudice began with Plato and Aristotle—the latter of whom believed knowledge should be deduced from first principles—and ended, to some extent, with the scientific revolution and Francis Bacon's idea that we must observe and manipulate nature in order to understand it, as was discussed in Chapter 1). Hacking argues theory does not always precede method. On the contrary, often observations come before theory. For Hacking, posing a question about the relative value of theory over method (or vice versa), or which precedes the other, is misleading 'because it treats theory as one rather uniform kind of thing and experiment as another' (ibid., 162).

BOX 4.1 ❖ WHICH CAME FIRST, EXPERIMENT OR THEORY?

Arno Penzias and R.W. Wilson, two radio astronomers working in Crawford Hill, New Jersey in the mid-1960s, were using a radio telescope to detect energy sources in the universe. They detected these sources and something else: a small amount of energy that seemed to be uniformly distributed everywhere in space. The scientists did what scientists typically do: assume that there was some kind of problem with their instrumentation and measurement, and set about trying to fix it. This included going to elaborate lengths to remove the pigeons (and their excrement) that nested on the radio telescope (Penzias and Wilson reasoned that the slight increase in temperature the radio telescope detected might be due to these pigeons or their droppings). Even after the scientists had dealt with all possible measuring and observing errors they could think of, there was still a uniformly distributed three degrees Kelvin difference in temperature between what they thought they should detect, and what they did detect. Here we have scientists who have observed and measured a phenomenon that is not described in the scientific literature. According to Thomas Kuhn, this would normally result in, well, nothing (see Chapter 2). Without a paradigm within which to fit this result, scientists would assume that there was some sort of measuring error that had yet to be detected.

What makes this an interesting story for sociologists is that at the same time that Penzias and Wilson were scraping pigeon droppings off their radio telescope, theoreticians at Princeton had just written an article speculating that if the Big Bang had created the universe, there would be a residual temperature from this initial explosion, and this energy would be detectable in the form of radio signals. And so, because Penzias and Wilson now had a (competing) theory within which to situate their research findings, they now had reportable research findings deserving further research. Another interesting aspect of this story is that Penzias and Wilson were awarded the Nobel Prize for their experimental work, rather than the Princeton theoreticians. Typically it is the other way around, since experiment is chiefly viewed as (only) either confirming or disputing theoretical research. See Bill Bryson's *A Short History of Nearly Everything* (2005).

How Sociologists Study Science and Technology

When sociologists study science, they are interested in how the scientific method works in practice, and how this practice has implications for broader questions about what constitutes knowledge (whether knowledge has to be entirely valid, or if it is sufficient for it to be practically valid, and so on). Some analyses argue what scientists do in practice bears little resemblance to the scientific method: indeed Paul Feyerabend argued so strongly against the idea that scientists use a uniform method for deriving knowledge that he called his book on the topic *Against Method* (1975) and Nancy Cartwright wrote a book called *How the Laws of Physics Lie* (1983), suggesting 'fundamental laws do not govern objects in reality; they govern only objects in models' (ibid., 18). The use of models in science complicates what we mean by reality and causation (see Chapter 2), and we focus on the scientific concept of models at a later point in this chapter.

These arguments are made on the basis that science is an inherently social process. And as a social process, it is necessarily *messy*. This is to say that behind the scientific method lies a cascade of related social paradigms and practices (Hacking, 1983). The challenge for the sociologist is to find out what these paradigms and practices are, and how they generate certain kinds of knowledge and not other kinds. Sociologists explore scientific practices through a variety of means, including studying scientists and studying scientific laboratory practices.

Studying Scientists

One way to study science as a social practice is to study the lives and work of scientists. A number of books provide biographies of scientists, such as *Einstein: His Life and Universe* (Isaacson, 2008) and *Never at Rest: A Biography of Isaac Newton* (Westfall, 1983). These biographies are sometimes written by scientists, and they can offer particular insights into the knowledge a particular scientist developed, as well as the disciplinary context in which she or he developed that knowledge. Some scientists write autobiographies, such as James Watson's *The Double Helix: A Personal Account of the Discovery of the Structure of DNA* (1968). These accounts provide fascinating insights into science as a social practice. For instance, Watson wrote his book in the wake of the controversy surrounding Watson, Crick, and Wilkins's use of Rosalind Franklin's scientific research to formulate the double-helix structure of DNA (see Box 5.1). The book reveals a defensive scientist who clearly denied the gender politics that permeated this scientific research.

Bruno Latour expands this genre by studying Pasteur and other scientists in *The Pasteurization of France* (1988). In keeping with Actor Network Theory (see Chapter 2), Latour presents an analysis of how various actants assembled allies that created particular knowledge about vaccinations (anthrax, in this case), public health, social policy, and public understandings of science. The history of science demonstrates a familiar tendency to simplify history by identifying a single scientist with great discoveries of knowledge, such as Newton and gravity, Einstein and energy, and Pasteur and vaccination. The picture Latour paints is very different.

Vaccination as a mainstay of national (in this case France) state policy resulted from a complex interaction between an array of different actants, according to Actor Network Theory, who assembled each other as allies. Latour points to the rise of hygienists as a category of experts whose mission was to clean up cities by providing clean running water and a drainage system for disposing of human and animal waste. In the rising field of hygiene, Pasteur represented a vital spokesperson. In Actor Network Theory language, Pasteur was an important ally. For Pasteur and other scientists working in laboratories on microbes, hygienists brought valuable publicity to their research program, which led to increased funding and fame for Pasteur. Physicians were concerned with treating sick people. For them, assembling hygienists and Pasteurians as allies made sense insofar as it furthered their agenda to rid the human population of disease.

And since Actor Network Theory seeks to explicitly engage with nonhuman actants, Latour focuses on bacteria as well. What made the bacteria with which the biologists worked into actants, according to Latour, was the series of laboratory trials Pasteur devised to transform 'useless matter' into 'full-blown substance' through classification as either pathogenic or domesticated (ibid., 35, 122): 'Pasteur and the ferment mutually exchange and enhance their properties, Pasteur helping the ferment show its mettle, the ferment "helping" Pasteur win one of his many medals' (ibid., 36). In other words, actants (Pasteur, lactic acid ferment, laboratory, microscope, hygienists, social policy, government regulation, etc.) gain and modify their definitions through the event of the experiment (ibid., 37).

So while this study ostensibly writes a history of a particular scientist, the growth of scientific knowledge and its transformation of society—'We are now nine billion because of Pasteur', writes Latour (ibid., 34)—what Latour actually does is demonstrate more of a mangle of disciplines (biology, hygiene, policy, family medicine), laboratories (microscopes, experiments), bacteria (cultured and uncultured), and populations (how the public came to understand vaccination through Pasteur's public demonstrations). Through analyses like these, sociologists demonstrate how science is social relations.

Studying Laboratories

Beginning in the late 1970s, sociologists began to turn their analytic gaze toward laboratories. Laboratories have a material culture, composed of special equipment, measuring techniques, and recording devices (Bauchspies, Croissant, and Restivo, 2006), as well as teams of researchers, technicians, cleaners, engineers, administrators, and so on. Laboratories use various symbols (the white laboratory coat, particular comic strips on the laboratory's door, blackboard, meeting room table), as well as a language, classification schemes, models, machines, and measuring techniques that serve to unite the scientists, and place them within a local social system (the laboratory), a wider social system (a university, an industrial corporation) and within wider society (Canada), and, as we have discussed (see Chapter 2), a paradigm (ibid.).

Laboratories involve people in particular social roles: from the lead researcher who earns the grants that pay the researchers' salaries to the researchers who organize and conduct experiments to the university administrators to whom the lead researcher reports, the graduate and undergraduate students who assist the researchers and earn their degrees through course and laboratory work, the technicians who keep the laboratory machines running, and the cleaners who attempt to maintain some semblance of cleanliness and order within the laboratory space.

Given sociologists maintain that science is social, the laboratory provides an ideal space to study the minutia of everyday scientific research. In the 1970s and 1980s, a number of books began to appear in the literature, including Latour and Woolgar's *Laboratory Life: The Social Construction of Scientific Facts* (1986), Knorr Cetina's *The Manufacture of Knowledge: An Essay on the Constructivist and Contextual Nature of Science* (1981), Pickering's *Constructing Quarks: A Sociological History of Particle Physics* (1991), Collins's *Changing Order: Replication and Induction in Scientific Practice* (1991), Lynch's *Art and Artifact in Laboratory Science: A Study of Shop Work and Shop Talk in a Research Laboratory* (1985), and Traweek's *Beamtimes and Lifetimes: The World of High Energy Physicists* (1986).

Several of these studies adopt an ethnographic approach. Ethnography is a qualitative method that uses participant observation, interviews, and archival data (maps, texts, and so on) to understand how a given society or culture operates (its logic, rationale, rituals, and structures). It is commonly used in anthropology and, to a somewhat lesser extent, in sociology. Since ethnographies were originally used by westerners to study non-westerners, some sociologists mimic this approach by considering the science laboratory as a foreign setting. The most well known example is Bruno Latour and Steve Woolgar's study of Roger Guillemin's laboratory at the Salk Institute. This study eventually became the canonical book *Laboratory Life* (1986).

The book adopts an explicit rhetorical technique in which Latour and Woolgar pretend they are absolute foreigners to science. They refer to the laboratory as a 'tribe' (ibid., 48) and they try to erase any previous knowledge they have of science and scientific practice by asking themselves the kind of questions anthropologists might ask of a newly discovered group of indigenous people. For example, when they witness a group of scientists arguing about an equation written on the laboratory blackboard, Latour and Woolgar write: 'Are the heated debates in front of the blackboard part of some gambling contest?' (ibid., 44).

Making Matters of Fact

Latour and Woolgar eventually settle on archival documents—publications, laboratory meeting reports, meeting minutes, blackboard scribbles, funding applications, and the like—as key indicators of what it is scientists actually *do* in laboratories. The scientists, they note, 'spend the greatest part of their day coding, marking, altering, correcting, reading, and writing' (ibid., 48–9). Everything scientists do appears to end up in some form of writing. Data gathered from laboratory discussions and so on are transcribed from spoken to written word. Machines—mass spectrometers, for instance—transcribe raw material into dots or lines on graphs. Through these transcriptions scientists produce matters of fact.

One of the main benefits of studying laboratories is that sociologists get to see how a scientific matter of fact is established *in situ*—as it happens in the setting, or what Latour refers to as *science in action* (1987). (Biologists, by the way, also distinguish between things they can manipulate *in situ*, like the body, and things they can only successfully manipulate *in vivo*, or outside of the body). And the main point sociologists make here is that *scientists do not use machines, calculations and so on to discover already-existing entities; rather, their own inscriptions produce entities.* Inscriptions, here, mean all the processes scientists go through to produce matters of fact; that is, how images produced by cameras attached to microscopes, smudges on graphs, and so on are made to mean particular things. For this reason, Ian Hacking (1983) argues phenomena are *created* in the laboratory, and Louis Pasteur famously said, 'In the fields of observation, chance favors only the prepared mind' (1854).

Using the example of the peptide TRF(H), Latour and Woolgar explore how Guillemin and his laboratory team inscribed this particular peptide. For instance, Guillemin limited the challenges his research would receive from other laboratories by restricting (i.e., inscribing) TRF(H) to its structure rather than physiology. Recognizing—as all good scientists do—that the best facts (defined as most convincing or most likely to be accepted by other scientists) are those divorced from the contingent circumstances that create them (tinkering with machines, restricting what counts as a particular object, changing the object itself so that it can be studied, etc.), scientists do their best to remove all of the contingent circumstances that go into producing a particular fact from their reports and publications.

Models and Measuring

One very important way in which scientists inscribe matters of fact is through the use of models. A model is a simplified version of reality. It takes into account certain phenomena, such as particular intervening variables, while acknowledging that at best it approximates reality. Hacking describes models as 'intermediaries' that 'siphon off some aspects of real phenomena, and connect them, by simplifying mathematical structures, to the theories that govern the phenomena' (1983: 217).

Models are often used when scientists cannot do experiments that take into account all variables. Climate change is a good example. Scientists

---

BOX 4.2 ☀ MAKING *PFIESTERIA PISCICIDA* A MATTER OF FACT

Pfiesteria piscicida is a tiny organism that lives parasitically in certain fish. Scientists are interested in this organism because it is thought to kill billions of fish in mid-Atlantic estuaries of the United States.

Astrid Schrader (2006) produced a wonderfully rich study of the laboratory life of this tiny organism. Her study demonstrates how inscriptions produce matters of fact. The scientists in the laboratory Schrader studied were interested in studying *P. piscicida*, and to do this, they needed to distinguish between what is naturally *P. piscicida* and what is environmentally induced during their life cycle. Defining this organism's life cycle itself requires classificatory exclusions and nearly limitless environmental interaction, not least of which is with symbiotic bacteria.

The interesting thing is that separating *P. piscicida* from their various metabolic and reproductive transformations (i.e., through *in vivo* culturing in laboratories) produces non-toxic *P. piscicida*. So, in classifying the organism, scientists *create an organism* that is not the object of their study. It is not that scientists cannot determine anything about *P. piscicida*: it is that their inscription produces a new entity that produces matters of fact. As Schrader puts it, 'how you get to know a species experimentally cannot be separated from the . . . question of what they are.' Different inscriptions establish different phenomena. See also Hannah Landecker's *Culturing Life: How Cells Became Technologies* (2007).

---

derive knowledge about climate change through observation and measurement of particular things such as polar ice melts. They also construct mathematical models that approximate certain features of Earth's atmosphere (such as chemical composition) and so on. As such, models rely on assumptions. As the number of assumptions increases, the validity of the model may decrease.

Another way in which scientists create matters of fact is through measuring and classifying. Measurement is a key feature of scientific practice, and scientists measure a diversity of things such as human fetal head circumference, planetary movement, and electron activity. Hacking analyzes the history of measuring to show that it did not really become formally systematized until scientists began to think that the world might be characterized by a set of numbers, or 'constants of nature' (ibid., 235). Newton's gravitational force G is an example.

Kuhn argued most measurement is part of normal science (see Chapter 2). That is, measurement does not typically jeopardize theory: it confirms theory, and sometimes calls for new technologies and puzzle solving. As such, Hacking argues measurement serves a function: scientists continue to measure—indeed they increasingly measure—because it focuses their research programs by mopping up anomalies. Scientists tend to refer to these anomalies as errors or artifacts. Making the connection between theory and methods, Hacking asks, 'Do measurements measure anything real in nature, or are they chiefly an artifact of the way in which we theorize?' (1983: 233).

Learning to See

Whether searching for quarks or distant solar systems, scientists rely heavily on technology, such as microscopes and telescopes, to see objects and their interactions. Images are part of daily scientific practice, as well as how scientists present their findings for dissemination. One of the most interesting aspects of learning to see is the difference between seeing and recognizing. A key task science students must learn is how to recognize objects. Sarah Franklin (2007), a sociologist who researched how and why scientists ostensibly cloned Dolly the sheep, discusses how the laboratory researchers helped Franklin to recognize DNA through a microscope. (One of the most interesting features of Franklin's work is that she showed the researchers did not actually clone Dolly; instead Dolly came from two distinct cell lines. The scientists demonstrated cell differentiation can be reversed, a feature not thought to be possible before this research was undertaken.) Only after many attempts, and scientists *telling her what to look for*, and *Franklin moving the slide around*, using different angles and so on, did Franklin learn to recognize DNA.

Hacking argues we do not actually see with a microscope. He quotes the President of the Royal Microscopical Society who stated:

> . . . microscopic vision is *sui generis* [unique, and not part of a larger concept of vision]. There is and there can be no comparison between microscopic and macroscopic vision. The images of minute objects are not delineated microscopically by means of the ordinary laws of refraction; they are not dioptical results, but depend entirely on the laws of diffraction. (in Hacking, 1983: 187)

Diffraction is used in optics to refer to the apparent bending of waves (light, sound, etc.) when they encounter small objects. When you look at a DVD or CD, the rainbow colours are a diffraction effect. Donna Haraway, a well-known philosopher of science, uses diffraction as a metaphor for alternate readings made possible by reading one thing through another. She writes: 'Diffraction does not produce "the same" displaced, as reflection and refraction do. Diffraction is a mapping of interference, not of replication, reflection, or reproduction. A diffraction pattern does not map where differences appear, but rather maps where the effects of difference appear' (Haraway, 1992; Barad, 2007).

Consider this explanation of the microscope from a well-known biology textbook, *Optical Methods in Biology*:

The microscopist can observe a familiar object in a low-power microscope and see a slightly enlarged image which is 'the same as' the object. Increase of magnification may reveal details in the object which are invisible to the naked eye; it is natural to assume that they, also, are 'the same as' the object. . . . But what is actually implied by the statement that 'the image is the same as the object?' Obviously the image is a purely optical effect. . . . The 'sameness' of object and image in fact implies that the physical interactions with the light beam that render the object visible to the eye (or which would render it visible, if large enough) are identical with those that lead to the formation of an image in the microscope. . . . Suppose however, that the radiation used to form the image is a beam of ultraviolet, X-rays, or electrons, or that the microscope employs some device which converts differences in phase to changes in intensity. The image then cannot possibly be 'the same' as the object, even in the limited sense just defined! The eye is unable to perceive ultraviolet, X-ray, or electron radiation, or to detect shifts of phase between light beams. . . . This line of thinking reveals that the image must be *a map of interactions between the specimen and the imaging radiation.* (in Hacking, 1983: 190)

Scientists know that microscopes do not reveal faithful images of objects. Jutta Schickore (2007) explores the construction and use of microscopes between 1740 and 1870. Microscopists—scientists, as opposed to rich people who used microscopes for entertainment purposes—originally distrusted microscopes and some banned them from their laboratories because they were viewed as distorting the perception of objects. In *The Microscope and the Eye*, Schickore examines how microscopes were variously understood to produce optical deceptions; provide accurate representations of objects; reveal the defects of the human eye; illustrate the need for expert observation and interpretation; and expose the relationship between observer, instrument (microscope), and observed (object), as well as to delineate what is possible to know about objects through sight alone.

The point for Hacking, Haraway, and Schickore is that microscopes neither sufficiently nor necessarily provide the truth about entities. They are part of what Pickering calls the mangle of practice, and as such they produce particular knowledge based on particular parameters (diffraction of light, light sensation against the retina, neurochemical signaling to the brain, the classification of objects, and so on). They also explicitly require, Hacking argues, interfering: 'The first lesson' any scientist learns about microscopes Hacking explains, is that we learn to 'see through a microscope by doing, not just by looking' (recall Sarah Franklin's experience, noted above) (1983: 189).

Learning how to recognize objects also reveals the social context of science. More experienced researchers teach less experienced researchers what to look for, what to record, and so on. Less experienced scientists tend to defer to more experienced scientists, particularly if the scientist is the established lead researcher funding the laboratory. Often this deferral is implicit, as individuals pick up on non-verbal cues and rhetorical tactics (such as sighing or emphasizing particular words). Further, recall the previous discussion of paradigms in Chapter 2, which argued normal science dictates the parameters of what objects, what relations, and what contexts will be studied. As Fleck wrote, 'to recognize a certain relation, many another relation must be misunderstood, denied, or overlooked' (1935: 30). Learning to see is as much about overlooking some observations as it is about concentrating on others.

Tinkering

Scientists also *make* objects through imaging. Michael Lynch and Steve Woolgar's *Representation in Scientific Practice* (1990) explores the ways in which images are produced through the use of certain angles, colours, contrasts, cropping, and resolution (see also Daston and Galison, 2007). Any study of laboratory life quickly reveals how much time scientists devote to getting the images they produce to correspond to what they need the images to resemble. Images are used extensively in scientific publications and are judged a crucial element of any generation of new scientific knowledge. So crucial are images that the leading scientific journal *Nature* has an ongoing 'image manipulation' discussion in its issues.

Janet Vertesi's PhD research in science and technology studies (2007) on the Mars Rover mission provides an excellent example. Vertesi examined the processes through which visual knowledge about Mars was accomplished. She found scientists engaged in ongoing discussions and debates about what the land surface of Mars should actually look like without the distortions produced by the Rover instrument. For these scientists, the issue became one of how to get the images produced by the Rover instrument to look like what scientists think Mars actually looks like, and then disseminate those images to the public. Undergraduate students were paid to sit at computers and physically alter the colour and contrasts of the images sent back to Earth from the Rover mission.

In other words, a fundamental aspect of scientific inscription involves tinkering as a way of getting material entities to behave in ways scientists want them to (Knorr Cetina, 1981). Pickering also takes up this concept of tinkering in his mangle approach to science. Pickering focuses on the ways in which physicists tinker with the machines they use. Using the example of building a Bubble Chamber, he demonstrates these machines often do not work in the ways the scientists and technicians want or expect them to. The machines, in the language of mangle theory, respond to the scientists' tinkering by producing expected results, producing no results at all, producing unexpected results, and so on. Other types of tinkering—with concepts, and with social relations inside and outside the laboratory—are a fundamental part of science as practice.

Learning how to see, tinkering, observation, and measurement are all part of tacit knowledge within science (Collins, 1974; Polanyi, 1958; Daston and Galison, 2007). Recall the discussion of Boyle and Hooke's air pump (see Chapters 1 and 2). Boyle's philosophical argument that science should be based on the consensus of reliable witnesses as to the phenomena produced by the air pump was difficult to sustain in the face of numerous problems people had in replicating both the air pump itself, and getting it to perform according to Boyle's instructions. Boyle relied on tacit knowledge about what worked and what did not work in the construction of the pump and getting it to function according to his aims. In other words, Boyle and Hooke developed a skill set in constructing and working the air pump. Boyle's contemporaries discovered that, without Boyle sharing this tacit knowledge in his directions on how to build and use the pump, the pump either did not work at all or did not work to the scientists' satisfaction. Scientists learn through trial and error that the descriptions of machines and experiments provided in official publications are not sufficient, and scientists often visit each other's laboratories to learn first-hand the skills necessary to make the equipment and/or experiment work.

A significant component of what science students learn in the laboratory is tacit knowledge. Sociologist Rebecca Scott (forthcoming) became aware of this when she enrolled in a summer workshop on placentas. The course had an extensive wet lab component, where students perfused placentas, extracted DNA, prepared placental tissue, and then observed it under the microscope. As a novice, Scott relied heavily on the skills the scientists around her had developed to get the objects to behave in ways they wanted them to, even when she understood the premise of the experiment and how the apparatus should produce the expected results. These scientists frequently pointed out the differences between what the textbook instructions said, and what they found works in practice.

Tacit knowledge demonstrates that expertise cannot be entirely formalized. For instance, describing his team's attempt to refine cell culturing, Ian Wilmut wrote:

> It seemed as if the problem was simply to find the right conditions for culture which would allow [embryonic sheep] ICM cells to multiply without differentiating. *Cell culture after all is a craft as much as it is a science; to a large extent improvements are made just by adding things and taking things away, and seeing what results.* The behavior of cells in culture can be modified by changing the conditions just as a gardener can influence the behavior of his plants. (in Franklin, 2007: 37, emphasis added)

By 'adding things and taking things away, and seeing what results' Wilmut draws attention to science as both a craft (tacit skill) and a formalized science.

Deploying Matters of Fact

Another important way in which sociologists study how science and technology work concerns the circulation of knowledge. Latour and Woolgar's work illustrates how scientists circulate knowledge within the laboratory—making marks on chalkboards, discussing what is being seen under the microscope and so on. As Lynch and Woolgar point out, science is a labour organization, and as such, has devised explicit and implicit rules about '*who* can know what, *who* is allowed to know, and *who* can say what they know' (1990). Taking the example of transforming an experiment into a paper for submission to an academic journal, laboratory teams have routinized ways of delegating responsibility for who gathers data, who writes what, who decides which journal to send the paper to, and, importantly, whose names appear on the paper and in what order. In the social sciences, most journal articles are single-authored, meaning knowledge tends to be generated by individual researchers, ostensibly working alone. In science the case is reversed. Most scientific articles have multiple authors; for example, it is not uncommon to find articles with 10 or more authors. In the social sciences, only people who actively participated in the research and write-up are supposed to have their name appear on the paper, and authors are listed in the order signifying who did the most work, to who did the least work (the lead author did the most work). As well, there is tacit knowledge about how to read these authorship lists. Again, in the sciences this is reversed. The last author listed tends to be the team leader, the most experienced and senior researcher, whose laboratory is funded through her/his grant money.

Writing papers for publication is a crucial stage in the scientific method because it is at this stage that the laboratory team officially reports their experimental design, data, and findings. Other researchers, worldwide, are able to scrutinize the design, data, and analysis through the peer-review process. The peer-review process is a key way in which scientists determine the relative reliability, validity, and worth of any given piece of research. Papers submitted for publication are sent out by the journal to other scientists known to be experts in the field. The process pivots on the reviews being blind (the reviewers know who the authors are, but the authors are not supposed to know who the reviewers are). As well, there is a clear hierarchy of journal prestige, of which scientists are keenly aware, and they devote considerable energy to crafting papers that will be accepted by top-tier journals.

Peter Medawar (1963) wrote a well-known article entitled 'Is the Scientific Paper a Fraud?'. The basis of Medawar's argument is that the process of transforming experimental work into a publishable paper is essentially a process of erasure. All of the tinkering with instruments, data, and objects; all of the inscriptions; power relations between scientists; hypotheses that go nowhere; false starts; and dead ends rarely make it into the paper. Indeed, writes Medawar, journal articles are explicitly written in a particular narrative style that not only describes a given experiment, but also makes a particular argument or set of arguments. In other words, the authors write the article in such a way as to lead the reader through the experiment as an event, toward the conclusion that the authors desire. Journal writing is an act of persuasion (much like philosophy).

Persuasion is a key motivating force in the advancement of knowledge. But it is crucial for the very practical reason of trying to secure funding. Scientists, especially senior scientists, devote a great deal of time to writing grant applications and participating in the review and adjudication of other researchers' grant applications. (I have lost count of the times that I have heard senior scientists lament that they barely have time to visit their own laboratories because they need to devote the bulk of their time to writing applications for funding). Research grants fund laboratories and all of the people and equipment that work in the laboratory. Without funding, there is virtually no empirical work in science. Scientists are amply aware of this and a significant part of the tacit knowledge we learned about earlier in this chapter concerns the ways in which scientists craft research proposals. To give one example, among health (basic and clinical) researchers in Canada, there is an informal consensus—discussed at laboratory coffee breaks—that scientists must actually have produced results on a particular topic before they have any chance of securing funding for the study. This presents a paradox for scientists: how to produce results without funding in order to secure funding to produce the results. Funding bodies have limited budgets and innovative research (more speculative by nature) constitutes a greater risk for funding payoff than research that only somewhat develops already well-established research. Senior researchers commonly advise junior researchers to forward their careers by working on established research.

Written reports in the form of journal articles are also written within what Fleck describes as a given thought collective (see Chapter 2). Thought collectives set the parameters of how scientists define phenomena, what questions scientists ask of phenomena in the first place, the kinds of research designs they employ to try to find answers to these questions, the ways in which the experiments themselves are carried out, and the ways in which all of this is transformed into a narrative account for dissemination to other researchers. As Bauchspies, Croissant, and Restivo point out, 'truth is a social achievement and can be contested . . . Consensus is achieved not because there is simply more evidence, but because an entire framework becomes accepted that fits into a larger worldview' (2006: 68–9).

Given sociology's emphasis on the relation between humans and non-humans in scientific practice (see Chapter 2), one of the salient features of what does not make it into formal scientific accounts of practice (journal articles, textbooks, etc.) is the agency of the nonhuman entities involved in scientific study. All of the mangle of practice, as Pickering calls it—the tinkering with machines and instruments, the bacteria that must be coaxed into different bacteria in order to be viably studied, the objects that do not respond in the ways the scientists want or expect—disappears in the final account. Whereas in *practice* all entities have agency, in official accounts only facts themselves have agency. Latour and Woolgar (1986) refer to this as inversion. Inversion is the process whereby researchers write their accounts such that all but the matters of fact disappear. Facts, in official accounts, are entirely responsible for their own establishment.

Summary

This chapter explored ways in which sociologists study what scientists do. The starting premise of sociological analyses is that science is a deeply social practice, dependent upon networks of human and nonhuman entities (Longino, 1990; Rouse, 1996). These entities come into contact with each other in specific contexts such as laboratories, the journal publishing industry, grant applications, team meetings, data collection, and so on. Sociologists use various methods to explore scientific practices. For instance, some sociologists such as Ian Hacking take an historical approach and use archival material to demonstrate how particular concepts such as measurement developed as a critical component of the scientific method. Other sociologists such as Bruno Latour study scientists working in laboratories to discern how matters of fact are created. Creating scientific facts are dependent upon a cascade of interrelated practices including learning to see, tinkering, modelling, and transcribing and inscribing data. Latour and others use a variety of techniques including ethnography, interviews, (participant) observation, and discourse analysis to study the life of the laboratory.

Of course, as sociologists use particular methods in our attempts to ascertain knowledge about how science works, even a modicum of reflexivity requires we acknowledge our own methods and theories are, like scientists', social practices through and through. Sociologists, like scientists, work from a number of assumptions and guiding principles. There is nothing inherent in sociology as a discipline that secures our access to valid and reliable accounts of scientific practice.

Sociologists who subscribe to positivism are generally on the same footing as scientists in attempting to derive true knowledge. Positivism is most strongly associated with Auguste Comte—sociology's founder—and Emile Durkheim. Comte argued sociology would arrive at a complete understanding of the universe through the positivist method and that 'from science comes prediction; from prediction comes action' (in Pickering, 1993: 622). Durkheim's (1895/1982) theory of social facts was also based on a positivist approach. Social facts, according to Durkheim, are all of the structures and norms that are external to the individual and that shape individual action.

Durkheim wanted to find correlations between social facts in order to reveal laws. By using positivist scientific methods, Durkheim argued, laws derived from social facts could—as far as possible—resemble laws referring to the material world. For sociologists, studying how scientists arrive at matters of fact from a positivist perspective means subscribing to the scientific method.

For sociologists studying scientific practices from standpoints other than positivism—critical or speculative realist, Actor Network Theory, Strong Programme, post-colonial, feminist, and so on—the challenge is to avoid reproducing positivist assumptions such as the aim of generating an account that is externally valid; that is, valid outside of the social, political, economic, and cultural context in which scientific practice *and* sociological practice take place. Sociologists must also contend with the various limitations of ethnography, discourse analysis, archival analysis, and other methods. These limitations, such as relativism, tend to incite sociologists to draw attention to the profound yet often unacknowledged relationship between science, technology, and power. This relationship is the subject of the next chapter.

Key Terms

Reliability   This is the consistency of a system or research finding when repeated under the same conditions.

Validity   This is a property of statements whereby the conclusions follow logically from the premises.

Scientific method   This is a set of techniques and procedures for acquiring new knowledge about phenomena.

Ethnography   This is a research method used primarily by social scientists to study a particular culture or group of people's rituals, organization, social system, development, and material circumstances.

Matter of fact   In science and technology studies this term refers to facts derived from science as an inherently social enterprise.

Archival research   This is a method of analyzing a variety of different kinds of documents including such things as digital recordings, books, and letters.

Science in action   This term is associated with Bruno Latour's book of the same name. It refers to the method of closely following scientists as they go about their day-to-day practice of conducting scientific research in order to analyze how science is practised in society.

Peer review   This is one of the ways in which disciplines self-regulate, ensure standards, and increase credibility. Peer-reviewed journals require that submitted papers be read and critiqued by a number of scholars in the same field. Only those papers that meet these scholars' standards will be published in the journal.

Critical Thinking Questions ···

1. Do sociologists make up facts? If so, how? Do sociologists use the same kinds of techniques—learning to see, tinkering, and measuring, for instance? What similarities and what differences are there in how sociology and science operate?

2. If another discipline were to study how sociology conducts research—how we know what we know—what would they find?

3. Does making up facts mean there is no such thing as facts?

4. Scientists know that only about one per cent of the bacteria in human drinking water can be cultured and that 99 per cent are viable but not culturable (VBNC). If we cannot test VBNC bacteria because they do not stay alive while being tested, how can we tell if they exist?

Suggested Readings ···

P. Doing, 'Give Me a Laboratory and I Will Raise a Discipline' in E. Hackett, O. Amsterdamska, M. Lynch, and J. Wajcman, eds, *The Handbook of Science and Technology Studies*, 3rd edn. Cambridge, MA: MIT Press, 2008, pp. 279–96. This article provides a synopsis and critique of Bruno Latour's well-known article (see below). The author criticizes Actor Network Theory for focusing on human actants despite its emphasis on taking nonhuman actants into account.

J. Dumit, *Picturing Personhood: Brain Scans and Biomedical Identity*. Princeton, NJ: Princeton University Press, 2004. This book provides a very interesting analysis of how brain scan technology, medicine, and culture work in tandem to inform our understanding of identity.

P. Galison, 'Bubble Chambers and the Experimental Workplace' in P. Achinstein and O. Hannaway, eds, *Observation, Experiment, and Hypothesis in Modern Physical Science*. Cambridge, MA: MIT Press, 1985, pp. 309–73. This wonderful chapter uses the example of bubble chamber laboratory research to demonstrate how science actually works in practice.

B. Latour, 'Give Me a Laboratory and I Will Raise the World' in Knorr Cetina and M. Mulkay, eds, *Science Observed: Perspectives on the Social Study of Science*. London: Sage, 1983, pp. 141–70. This well-known article makes the case that the scientific laboratory is a sort of microcosm of science and technology, and as such, when sociologists study laboratories, they are analyzing much broader ways of understanding our world, including politics, economics, geography, and culture.

S. Shapin, 'History of Science and its Sociological Reconstruction', *History of Science*, 20 (1982): 157–211. This important article outlines how the fields of the history of science and science and technology studies overlap and inform each other. It argues that sociologists have reconstructed scientific events of the past in order to emphasize issues of power and inequality.

Websites and Films

Bruno Latour's website
www.bruno-latour.fr
> Bruno Latour is a well-known sociologist and creator of Actor Network Theory. His website contains details of his publications, current and upcoming projects, and many additional links.

Janet Vertesi's website
http://janet.vertesi.com
> Janet Vertesi conducts very interesting research on NASA's Mars missions. Her research analyzes the inherently social nature of the photographic images the Mars robots transmit to earth.

NASA's Rover Mission
http://marsrover.nasa.gov/home/index.html
> This website provides a wealth of images and other information about the Mars missions.

Monty Python's interpretation of the scientific method
www.youtube.com/watch?v=k2MhMsLn9B0
> This humourous spoof critiques the deduction and induction principles of the scientific method.

The Quest
> This National Film Board of Canada film recounts the story of Frederick Banting, researcher and scientist at the University of Toronto, who produced insulin. The film is interesting from a sociology of science perspective because it depicts scientific practice as the day-to-day activities of scientists in their laboratories.

5   Science and/as Power

Learning Objectives

In this chapter we learn:
- 🌐 Power circulates throughout science and technology in particular ways, including research agendas, funding, laboratory practices, and publishing;
- 🌐 Political economy studies analyze the association between science, technology, and the development of capitalism and the market economy;
- 🌐 Feminist science studies analyze how science is used to shape political and cultural claims about sex and gender;
- 🌐 Post-colonial science studies analyze how science is used to shape political and cultural claims about race and ethnicity;
- 🌐 Disability science studies analyze how science and technology are used to shape political and cultural claims about people with disabilities.

Introduction

Sociological analyses of science and technology begin from the premise that science has never been pure. This, as Chapter 1 details, means sociologists analyze historical and current scientific practices within the social, political, economic, and cultural contexts in which they operate. Sociology is defined as studying the logic (logos) of society (socio), and science and technology play a vital and central role in how western and non-western cultures define themselves. Chapter 2 discussed the major theoretical tools sociologists bring to bear on scientific practice. Sociologists maintain science and technology are thoroughly social.

The aim of this chapter is to focus on science and technology and power. Power is a central concept in sociology. All sociological analyses are, at least to some degree, typically focused on power as an organizing force in the formation and maintenance of social structures, institutions, and practices. This chapter considers science and technology as social practices in relation to four central topics within the discipline of sociology as a whole: the political economy of society, sex/gender, race/ethnicity, and (dis)ability. Following the first four chapters' emphasis on science and technology as social relations, this chapter will explore political economy analyses of science and technology and focus attention on science and technology as products of the historical development of capitalism and the market economy, wage labour, colonialism, international trade, and so on.

Sociologists are also centrally concerned with the ways in which science and technology have been deployed within cultures to define and classify sex, race, and ability. Indeed, the very separation of sex and gender, race and ethnicity, and ability and disability are historically entangled with scientific research. Sociologists are primarily concerned to analyze the uneven distribution of resources and opportunities that follow the categorization of particular groups within society, as well as the implications for peoples' lived experiences.

Political Economies of Science and Technology

The Enlightenment was not a naive project to emancipate humankind from ignorance and oppression. Enlightenment philosophers, politicians, and economists debated and conceived of a new social order, but this new structure did not concentrate on enfranchising the common labourer. The Enlightenment's attitude towards nature was one of domination and control. Indeed, the Enlightenment's focus on rationality became allied with domination. As we saw in Chapter 1, Kant's disquiet about the Lisbon earthquake and its aftermath led to the development of his influential philosophy, which provided a rationality for the separation of humans (and human consciousness) from the universe. As Aronowitz observes:

> . . . the floods, torrential rains, snows, earthquakes, as much as disease, in short, the unexpected 'revolts' of nature have become phenomena to be controlled; these events define the task of science, the horizon in the quest for domination. Just as nature is understood as subject to subsumption under human powers, so humans themselves are increasingly regarded as controllable. (1988: 318)

For sociologists, modern science and technology are inextricably implicated in the overarching goal of better understanding, and thereby controlling, human labour and capital. We see this most clearly, perhaps, in the uneven development of scientific research whereby more investment is made into research that benefits military and industrial elites (Hess, 2009). As Chapter 1 outlined, research within Canadian universities today is increasingly moving away from public funding to private sources, with a concomitant emphasis on applied science, technology transfer, and economic competitiveness (Frickel et al., 2010). Just by virtue of Big Science's monetary, infrastructure, and bureaucratic requirements, scientific inquiry is now largely subject to corporate and state purposes (Aronowitz, 1988). As Frickel et al. argue, 'Because elites set agendas for both public and private funding sources, and because scientific research is increasingly complex, technology-laden, and expensive, there is a systematic tendency for knowledge production to rest on the cultural assumptions and material interests of privileged groups' (2010: 446).

Decisions about what knowledge should be pursued and what should remain off limits to scientists tend to occur amongst elites including federal legislators. Industry and state funding priorities reflect the economic, political, and social programs of these elites. These funding priorities often concentrate on the effects rather than causes of social problems. Heart disease, stroke, and obesity are proper objects of health research; the work environment and poverty are not. This demarcation of the object of study has much to do with large-scale pharmaceutical funding linked to hospital funding, governments, etc. As such, Remington observes this reorientation:

> ... measures the value of science almost exclusively in terms of utility and profitability for the national economy, military preparedness, corporate competitiveness and medical efficiency. Conspicuous by their absence are rationales of science (a) as a means of addressing social and environmental problems in open and pluralistic ways, and (b) as a cultural institution with an integrity of its own. . . . (1988: 56)

Science, technology, and power are co-extensive. Exploring the political economies of science and technology will, therefore, provide further collaboration of the axiom that science and technology are social relations.

As Fleck, Kuhn, Bloor, Latour, and others point out, science has become normative from the outset of any given research program that must specify the object of its study within already established classificatory systems, through the theories and methods applied to the object, and to the form and content of the results. We know from Chapter 3 that, after quantum theory, science and technology are interventions into a given field of study. The laws of science describe 'the relation of humans to the object of knowledge, not the objects themselves, taken at a distance' (Aronowitz, 1988: 331). As such, 'natural objects are also socially constructed' (ibid., 344). For this reason, as we learned in Chapter 2, Fleck argues facts are produced rather than discovered.

Explicit control of science and scientists by the state is unnecessary; these norms have already been internalized within the structures and practice of science. The principal investigator or lead scientist in the university laboratory setting is essentially a manager who must raise funds; hire and fire academic, technical, and administrative staff; and administer the relationship of the laboratory (and all of its labourers) with the university administration, funding bodies, and the scientific community (Aronowitz, 1988). Scientists have become, of necessity, entrepreneurs (Remington, 1988). Moreover, senior scientists must respond to what Remington terms the 'binding of inquiry' whereby 'scientific inquiry and its values and assumptions are bound to a highly utilitarian conception of knowledge, to "rational" institutional policy goals, to the technological infrastructure and to the framework of a politicized reward system for scientists' (1988: 62–3).

Graduate students aspiring to gain membership within academia must participate in these normative structures and practices (including participating in funded research and selling their labour). Science and technology are labour processes, and the division of labour is made between administrators, academics (scientists), and manual labourers. The scientific community defines and regulates entrance into and progression through its system, thus legitimating the conventions it uses through the reproduction of each new generation of scientists, who become necessarily implicated in the ongoing normatization of science. Part of this normatization involves subscribing to the ideal of modern science, that science itself provides the best (and only) source of valid and reliable knowledge about the world. Through this technique of demarcation—separating science from every other pursuit of knowledge—science increases its power and domination.

Feminist Science Studies

The field of feminist science studies has mushroomed since the 1980 publication of Carolyn Merchant's influential book, *The Death of Nature*. This section concentrates on major feminist approaches to science and technology. For the most part, feminist science studies critiques science and technology—what I have elsewhere termed the project of uncovering the *culture of matter* (Hird, 2003)—using social constructionism as its primary theoretical tool. Feminists largely direct their efforts to critiques of science because theories about the supposed nature of sex differences emerged within political, economic, and social discourses during the eighteenth century and more recently (social Darwinism and sociobiology, for example) deployed science as a key source of evidence for 'solutions to increasing questions about sexual and racial equality' (Schiebinger, 1993: 9). This scientific research cohered around an emphasis on differences between women and men which informed and was informed by political, economic, and social movements interested in maintaining institutions and structures (including the nuclear family, schools, the paid and unpaid labour force, immigration policy, and so on) that enfranchised men and disenfranchised women. The development of technologies such as the typewriter, telephone, clothes washer, and electric stove are also steeped in gender relations. Feminist science studies analyze the historical development of the tandem association of these technologies with repetitive, unskilled labour and women.

Women as Producers of Scientific Knowledge

Let's take this short quiz: how many Canadians do you think have heard of Albert Einstein, David Suzuki, and Stephen Hawking? How many Canadians have heard of Henrietta Leavitt, Annie Cannon, or Barbara McClintock?

Leavitt invented a method—the Standard Candle—to determine the relative distances between Cepheids (a type of star). In the late 1800s, Cannon devised a classification system for stars, which was of such importance that it is still used today. McClintock won the Nobel Prize in 1983 for discovering genetic transposons (see Box 3.3).

A number of texts (Longino, 1990; Mayberry, Subramaniam, and Weasel 2001; Small, 1984; Rossiter, 1984) examine women in science, arguing female scientists have had to particularly struggle against male-dominated disciplines within the natural sciences. Some analyses (Ainley, 1990) survey statistical trends of female students and academics in various science disciplines. Other analyses are interested in exploring the working lives of women biologists (Keller, 1983; Lancaster, 1989, 1991); primatologists (Altmann,

---

BOX 5.1 ☀ ROSALIND FRANKLIN

Rosalind Franklin (1920–58) was a British scientist who completed path-breaking work at King's College, London. She was a physicist, chemist, biologist, and X-ray crystallographer. Although she made important contributions to our understanding of tobacco mosaic and polio viruses, she is best known for her X-ray diffraction images that revealed the double-helix structure of the gene.

James Watson and Francis Crick used Franklin's 'photograph 51', as it has become famously known, in their research on the double-helix structure of DNA. Watson and Crick were given Franklin's images by Maurice Wilkins. He got them from one of Franklin's graduate students at the direction of Franklin's boss, John Randall. Franklin's images were used without her permission or knowledge. Max Perutz also gave Crick a report that Franklin wrote for a medical research council that contained many of her careful molecular model calculations. Her photographs and calculations provided invaluable clues for Watson and Crick's research.

Watson, Crick, and Wilkins received the Nobel Prize for work that relied on Franklin's contributions. Franklin died before the Prize was awarded, and thus never received this recognition. While Wilkins, Perutz, and Crick later acknowledged a certain level of sexism that Franklin had to endure as a female scientist, and the value of her work in Watson and Crick's research, this only occurred because of criticisms they sustained over their treatment of Franklin and their initial lack of acknowledgment of her vital contributions.

Feminist science studies scholars are interested in the ways in which women scientists continue to endure sexism in the workplace, and how work produced by women scientists continues to be defined in ways that differ from the work produced by men scientists. For more information watch *Secret of Photo 51* (PBS NOVA).

1980; Fedigan, 1984; Haraway, 1989; Hrdy, 1974, 1981, 1986, 1997; Rowell, 1974, 1979, 1984; Small, 1984; Zihlman, 1985); entomologists and astronomers (Schiebinger, 1989); mathematicians (Henrion, 1997); physicists (Barad, 2001; Keller, 1977; Wertheim, 1995) and engineers (Meilwee and Robinson, 1992). These diverse studies emphasize how women affect, and are affected by, the largely male culture of science. And indeed it is partially thanks to the feminist movement that women have gained entry into the fields of science and technology at all.

Women and Knowledge

A second area of feminist science studies concerns women's ways of knowing science. Some research argues women have historically been marginalized from all processes of scientific endeavour, which, when coupled with the thesis that women view and engage with the world differently than men, prompts female scientists to approach scientific questions from a less mainstream perspective, challenging fundamental assumptions about issues such as objectivity. For instance, Evelyn Fox Keller (1983) suggests Barbara McClintock's Nobel Prize–winning research on genetic transposons (see Box 3.3) was produced by her greater openness to alternative accounts of genetics because her greater exclusion from male science also meant she was less indoctrinated with their norms.

The argument that women approach nature and science questions from a fundamentally different perspective is particularly distilled in theories of science as social knowledge, and raises the question of the possibility of a distinct female way of knowing science (see Hankinson-Nelson and Nelson, 1996; Hubbard, 1989; Longino, 1990; Schiebinger, 1989; Stengers, 1997, 2000). Again, these analyses emphasize the social, political, and economic features of women in science rather than the actual objects that female scientists study.

Women and the Environment

Adopting many of the premises of women and knowledge, a third area feminists theorize is eco-feminism (for a useful summary see Soper, 1995). Eco-feminism comprises a diverse range of approaches, both theoretical and practical, in response to the impact of humans on the environment. One area that eco-feminist scholarship focuses on is science's construction of nature and the environment as things that can (and should) be managed (read: dominated) by humans. As such, some feminist scholars focus on how the commodification of nature has tended to work in tandem with other forms of oppression such as colonial and capitalist exploitation (Mortimer-Sandilands and Erickson, 2010; Di Chiro, 1998; Murphy, 2006). These analyses bring nature and the environment into discussions of social justice, pointing out that race and gender are often connected in scientific, political, and philosophical arguments about the inferiority of people who have been classified into groups (such as women, racialized people, and so on).

Women and Technology

A fourth area of feminist attention is the relation between women and technology. Analyses within this field tend to focus most often on reproductive technologies and cyber cultures. Many feminist considerations of the future of sexual difference focus on the impact reproductive technologies might have on conceptualizations of women's bodies. A host of feminist contributions concentrate on reproductive technologies (and increasingly genetics, see Franklin, 1995, 2000) as a form of female exploitation that reveals significant race and class privileges. Feminists, for instance, draw attention to the long history of testing female birth control hormones on women from developing countries and the ways in which women become implicated—through reproductive technologies—in societal ideals about having particular kinds of children (Briggs, 2002; Murphy, 1989; Overall, 1989; Sawicki, 1999; Spallone and Steinberg, 1987; Weir, 1998; see Donchin, 1989 for a useful review). Others see transformative possibilities, for instance through cyborgean techno-bodies, that technology potentially offers (Lykke and Braidotti, 1996; Broadhurst Dixon and Cassidy, 1998; Featherstone and Burrows, 1995; Gray, 1995; Haraway, 1991; Johnson, 1999; Plant, 1997). Still others—recall, for example, Ursula Franklin's critique of how the introduction of typewriters led to the increased exploitation of women—critiqued the gendered production and consumption of technology, particularly as these relate to labour.

Sex Differences and the History of Science

The fifth area of feminist analysis concerns the historical development of western science—mostly concentrating on the seventeenth, eighteenth, and nineteenth centuries—which enabled western social, economic, and cultural discourses to emphasize sex differences rather than sex diversity, intrasex (within sex) differences, and intersex (between sex) similarities (Hyde, 2005). Some analyses focus on the social construction of scientific knowledge concerned with the essence of sex and sex differences: gonads, hormones, chromosomes, and genes (Hird, 2004b). These analyses critique scientific accounts of sexual difference based upon skeletons, hormones, chromosomes, egg and sperm activity, and animal behaviour.

For instance, Emily Martin's (1990) analysis of the inscription of cultural notions of sex differences on to the physical processes of egg and sperm activity provides the basis for an analysis of human gonads as an important site deployed in the social construction of sexual difference. Martin's research, looking at how egg and sperm are described in medical textbooks, shows that sperm are attributed with ideal masculine characteristics, including strength, tenacity, advanced reasoning, and so on. Eggs are ascribed ideal feminine characteristics that emphasize passivity. These descriptions run counter to actual research on sperm and eggs, which show that eggs are much larger than sperm, that sperm movement is fairly haphazard, and that sperms' attempts to escape from the egg are thwarted by the egg's ability to force sperm to attach to the egg's surface and then to dissolve the outer layers of the sperm. As Anne Fausto-Sterling suggests, behind debates about sexual reproduction 'lurk some heavy-duty social questions about sex, gender, power, and the social structure of European culture. . . . In the work of the established evolutionary biologists, past and present, talking about eggs and sperm gives us permission to prescribe appropriate gender behaviors' (1997: 54, 57).

In another example, Nellie Oudshoorn (1994) provides an important analysis of the development of research on hormones, which were 'defined' to emphasize sex differences. However successful feminist arguments concerning the social construction of sex differences have been within academia and the public in general, there remains a persistent and robust recourse to a biological notion of sexual difference based upon often cursory notions of testosterone levels or X and Y chromosomes (Roberts, 2007).

The Separation of Nature and Culture

Western science, as these analyses demonstrate, has been and continues to be a powerful and influential shaper of knowledge about sex differences. The concept of sex differences, as these analyses also show, is dependent upon Kant's legacy of separating humans (culture) from nature (see Chapters 1 and 2). A number of feminist science studies scholars challenge this bifurcation through superb analyses. The most well known is famed biologist and social theorist Donna Haraway (1989, 1991, 1997, 2008). Haraway uses a 'natureculture' approach to detail the indelible and complex entanglement of nature/culture through which she weaves a series of connected stories about kinship, politics, sociality, economics, and environments. Haraway explicitly engages with a host of human and nonhuman entities (dogs, microscopes, animal breeding organizations, pharmaceutical industry copyright laws, immigration, policing policy, etc.) to think about how humans might better engage ethically with each other and with other species.

Elizabeth Wilson's innovative feminist engagement with science provides another excellent example. The central tenet of her recent book, *Psychosomatic: Feminism and the Neurological Body* (2004), is that the physical body (soma) and the mind (psyche) do not correspond to different realities of the body. Wilson begins by arguing that Sigmund Freud's early work on neurology did not separate the mind from the body, as his later analyses would. The model that Freud develops (nerves-penis-cortex-psyche) relies upon each element operating within 'circuitous relations' rather than as separate elements in 'determinable relations' (2004: 19). The mind and body, in other words, co-constitute each other, or 'make each other up'. Wilson builds on Freud's decisive work through an analysis of gastrointestinal complaints. Wilson traces the development of both biological and psychoanalytic theories of gastrointestinal disorders such that both eventually sever their relationship with the other: each constructs theories of the gut that obviate all connection between body and mind in such a way as to foreclose explorations of the ways in which the gut might co-constitute both soma and psyche:

> Maybe ingestion and digestion aren't just metaphors for internalization; perhaps they are 'actual' mechanisms for relating to others. That is, perhaps gut pathology doesn't stand in for ideational disruption, but is another form of perturbed relations to others—a form that is enacted enterologically. The large stocks of serotonin in the gut, the morphological similarities between gut neurons and brain neurons, and the clearly psychological character of gut function all suggest that it is not just ideation that is disrupted in depression; it is also the gut. (Wilson, 2004: 45)

In effect, what Wilson provides us with is an innovative way to think about issues of concern to feminist researchers including eating disorders such as anorexia nervosa and bulimia, depression, hormone replacement therapy, pregnancy, birthing, and breastfeeding—any issue that concerns gender, the body, and society.

Racialized, Indigenous, and Anti-racist Science Studies

Science studies and indigenous studies intersect on a number of issues. Sociologists analyze how scientific research is implicated in our changing understanding of what it means to be a living organism, what it means to be human, and what it means to be sexed. Scientific research is equally implicated in our changing understanding of what it means to be indigenous and racialized.

Science and technology have, as sociologists point out, an ambivalent relationship with questions of race and ethnicity. On the one hand, scientific research has been used to structure and/or inform political agendas, such as those of the Nazi Party, which sought the extermination of particular groups of people it deemed unfit (less intelligent, physically weaker, and so on), including Jews and other ethnic groups, homosexuals, people with disabilities, and others. Scientists and social science critics of science have also been at the forefront of challenging both the science behind this sort of political agenda, as well as the political agenda itself. For instance, one of the most well known examples of this is evolutionary biologist Stephen J. Gould's famous book *The*

BOX 5.2 ☼ EUGENICS IN CANADA

Eugenics arguments have been around a long time, and are most strongly associated with the Nazi Party's final solution, which sought the extermination of all Jews, people with disabilities, travellers (gypsies), homosexuals, and others. But eugenics runs through the history of Canadian society.

Consider *The Famous Five* or the *Valiant Five*, as they are alternatively known. The terms refer to a group of five prominent women—Emily Murphy, Henrietta Muir Edwards, Nellie McClung, Louise McKinney, and Irene Parlby—who, in 1927, challenged the Supreme Court of Canada to define whether women counted as 'persons' eligible to be Senators in the Canadian government. This challenge became known as the Persons Case. The Supreme Court ruled that women were not persons. Only when the women appealed the case to the Judicial Committee of the British Privy Council did they win. Emily Murphy became the first woman magistrate in Canada. Henrietta Edwards was a founding member of the Victorian Order of Nurses, Nellie McClung was a member of the Alberta legislature, Louise McKinney was the first woman to be elected to the Legislative Assembly in Alberta (and anywhere in Canada and, indeed, the whole of the British Empire), and Irene Parlby became the first female Cabinet Minister in Alberta.

Emily Murphy and Nellie McClung were also against Canadian immigration, and they championed eugenics. Both were instrumental in bringing into legislation the Sexual Sterilization Act of Alberta (which was only repealed in 1971). Under the authority of this Act, thousands of Albertans were sterilized without their own knowledge (Maclaren, 1990).

Mismeasure of Man (1996) that painstakingly critiques the scientific basis of the association between intelligence and race.

In this section, we examine the complicated and ambivalent relationship between indigenous studies and scientific research. Racialized groups have at times embraced scientific research concerned with genetic markers that can be used to inform understandings of ethnic group membership and specific health concerns (Epstein, 2007). On the other hand, racialized groups are wary of the ways in which science can equally be used to further stigmatize, survey, and oppress already colonized peoples.

Re-thinking the Biology of Race

The designation of human beings according to specific phenotypic characteristics—skin colour and facial features, for example—has been an important part of society's structuring processes. In other words, race has been, and continues to be, one of the most enduring ways that any given society defines and structures itself (see Hacking, 2005).

Canadian society is ostensibly structured—written into our laws and policies—around the concept of multiculturalism rather than race, per se. This means, officially, the Canadian government recognizes a diversity of peoples, and this diversity explicitly shapes government policy. Note the use of the term multi*cultural* to designate a combination of ethnicity and culture. The second edition of Vic Satzewich and Nikolaos Liodakis's book *'Race' and Ethnicity in Canada* (2010) provides an excellent introduction to the ways in which Canadian society has understood the concepts of race, ethnicity, and multiculturalism in different historical contexts, such as waves of immigration; changing French, English and Aboriginal relations; and the Second World War.

The Canadian government's focus on multiculturalism as opposed to race marks a central development in both scientific and social scientific approaches to race. The American Anthropological Association's (AAA) *Statement on Race* (1998) is exemplary of the academic community's critique of the concept of race. The Statement argues people cannot be easily or clearly categorized into particular biologically based groups, that the overwhelming majority of physical variations occur within racially designated groups, and that since humans share genetic material through sexual reproduction across racialized groups, humankind is a single race with more intra- (*within* group) differences than inter- (*between*) group differences. Moreover, the diversity of phenotypes (observable characteristics) and their relation to each other (for instance, between skin colour and curly or straight hair) precludes any kind of easy or valid association of phenotypes with particular races. The point is that the concept of race has always referred much more to the social meanings societies have attributed to phenotypes as a basis for discrimination against particular groups in society. Indeed the AAA statement makes the link between race and oppressions such as slavery explicit:

From its inception, this modern concept of 'race' was modeled after an ancient theorem of the Great Chain of Being, which posited natural categories on a hierarchy established by God or nature. Thus 'race' was a mode of classification linked specifically to peoples in the colonial situation. It subsumed a growing ideology of inequality devised to rationalize European attitudes and treatment of the conquered and enslaved peoples. Proponents of slavery in particular during the 19th century used 'race' to justify the retention of slavery. The ideology magnified the differences among Europeans, Africans, and Indians, established a rigid hierarchy of socially exclusive categories, underscored and bolstered unequal rank and status differences, and provided the rationalization that the inequality was natural or God-given. The different physical traits of African-Americans and Indians became markers or symbols of their status differences. (AAA, 1998)

Box 5.3 ❋ Lander's 99.9 per cent rule

Eric Lander, a biology professor at MIT, wrote a famous chapter on statistics and genome mapping. In this chapter, Lander outlines the difficult challenges scientists face when trying to determine a genetic basis for inherited diseases. It is more difficult, writes Lander, than finding a needle in a haystack: a two-gram needle in a 6,000-kilogram haystack is 1,000 times easier to find than finding a genetic 'mistake' in just over one part in 10^{10} base pairs of DNA inherited from either parent.

The human genome has 23 chromosome pairs (one pair of sex chromosomes and 22 pairs of autosomes). The typical gene is estimated to be about 30,000 nucleotides long. *If two genomes were randomly chosen from the human population, they would be about 99.9 per cent identical.* This high degree of sameness confirms our common heritage as a species. However, the 0.1 per cent difference translates to about three million sequence differences, so each individual human is unique.

Lander details genetic mapping and physical mapping as two techniques geneticists use to try to identify genes associated with inherited diseases. Social scientists, policy-makers, and politicians have picked up on Lander's 99.9 per cent rule to argue that race is not valid. Sociologists further note the myriad decisions that have to be made by geneticists about what they are going to search for and how, and that these decisions are based on statistical parameters and scientific conventions (i.e., genes do not in themselves reveal inherited diseases). See E.S. Lander (1995).

'Race', writes Arun Saldanha, 'refers then to the cultural *representation* of people, not to people themselves' (2006: 9). This social constructionist approach to race is called race pragmatism (Nelson, 2008: 760). Race pragmatists focus on the historical oppressions and future risks of scientific research on race, and we will read more about these critiques in a moment.

There are some interesting caveats to this approach, however, coming from science studies. Nelson identifies these approaches as race naturalism (2008: 760). Howard Winant (2004) and Naomi Zack (2002), for instance, argue phenotypic characteristics do exist, but that their meaning is created and sustained by society (through culture, economics, politics, and law). Ian Hacking takes a stronger position, stating '[n]ature makes differences between individuals. These differences are real, not constructed' (2005: 103). Arun Saldanha (2006) argues social scientists' immediate move to race as representation belies an anxiety about dealing with matter generally, and we must learn strategies to critically engage with race's biological aspect (Tyler, 2009).

Moreover, racialized and indigenous individuals and groups themselves sometimes endorse race as a biological concept. Alondra Nelson's (2008) ethnographic study of African American and black British individuals' use of genetic genealogy testing to better understand and narrate their own ancestral histories acknowledges racialized groups themselves sustain an ambivalent relationship with race naturalism. At different times and in different contexts, racialized groups selectively use genetics research to inform claims making around a number of issues including group membership, health, and so on. So, for instance, while Nelson's research respondents sought genetic testing to provide further clues to their ancestry, they *also* engaged in 'highly situated objective and affiliative' self-fashioning, interpreting genetic test results in the context of their 'genealogical aspirations' (2008: 759; see also Condit et al., 2004).

Race, Genes, and Health

As Dorothy Roberts (2008) points out, while race might not be considered a viable concept amongst most sociologists and scientists, it gains purchase within society through politics, culture, and most recently, through genetics (see also Hartigan, 2008). As with so many social issues, mapping the human genome is represented as both a good and bad thing. Some researchers point to the ways in which scientists' better understanding of the human genome may positively contribute to health research. Human genetics studies the inheritance of characteristics from one generation to the next, including a number of genetic conditions and diseases. We now know of more than 11,000 conditions that result from single gene alterations (Paddaiah, 2008). For instance, Down's Syndrome is a consequence of one extra chromosome. Cystic fibrosis, Duchenne muscular dystrophy, sickle cell disease, and hemophilia are consequences of a single gene alteration. Other conditions are due to complex interactions between genes and environment, including heart disease, diabetes, and a cleft lip or palate.

Paddaiah (2008) argues the study of human genetics can help some of the most marginalized, such as tribal populations. Understanding the connection between race and genetic conditions affecting health, Paddaiah suggests, may lead to people's more informed decision-making. For instance, human genetics has discovered that when people with hemoglobin AS have children, one-quarter of those children will eventually die from anemia. People who are hemoglobin AS can now be informed of their condition and of the possible consequences of producing children with partners who are also affected by this condition. In another example, anti-malarial drugs administered during the Second World War to US soldiers revealed about 10 per cent of black soldiers and a smaller number of white soldiers of Mediterranean origin developed severe anemia.

Racialized groups have themselves lobbied governments to make health research race-specific through the inclusion of racialized groups in health research sampling for the purposes of determining the best specific treatment options for racialized people (Epstein, 2007). Up until the 1990s health researchers mainly researched white, male, middle-class people and either assumed or argued that research findings generated from this population were generalizable to everyone (including women, children, elderly, and racialized peoples). To take just one example, researchers studying the relationship between obesity and cancer of the breast and uterus, had undertaken a pilot study using only men (Jaschik, 1990). In the United States, it took an act of congress (the Revitalization Act was passed in 1993) and years of lobbying on the part of many groups to change the National Institutes of Health policies to include anyone other than white men as subjects in clinical trials.

Steven Epstein's book *Inclusion: The Politics of Difference in Medical Research* (2007) details the logic of what he calls the 'inclusion-and-difference' paradigm within health research. The term refers to a set of arguments made by diverse (and sometimes competing) groups including clinical researchers, policy-makers, lobbyists, human rights groups, and politicians who sought the inclusion of members of particular groups—mainly women and racialized people, but to some extent also children and people living in poverty—who have traditionally been underrepresented in clinical research, and the measurement across these groups with regard to clinical outcomes (how effective a treatment is, for instance). Epstein shows that the inclusion-and-difference paradigm was largely successful in the United States in the sense that clinical researchers were obliged through legislation and funding policy to include underrepresented groups. He questions, however, the implications of making compulsory research that focuses on categorizing people and then finding differences between these categories. Since a great deal of human rights battles have been won by arguing against the categorization of people, the major criticism of the inclusion-and-difference paradigm is that it will reinforce assumptions about differences between women and men, and between racialized and white people. When research policies, clinical practices, government funding, and the entire production of clinical research is based upon the supposition that differences *will* be found between groups of people, then it's a pretty safe bet that clinical research will produce findings that reveal differences.

The debates within the inclusion-and-difference paradigm show that the ongoing push to link genetics, race, and health is controversial. Genetics researchers and epidemiologists (epidemiology is the study of factors that affect the health of a population) acknowledge that the environment plays a significant role in determining health. Well-known epidemiologist Richard Cooper (2003), for instance, argues the racial variance of the incidence (how frequent the development of a new illness is in a population) and prevalence (the number of cases of an illness in a population) of many diseases in the

United States is most likely due to social, cultural, economic, and political factors because poorer minority populations in the US have less access to, and poorer quality of, care. Indeed, Cooper argues the concept of race 'is and always has been in essence a social construct' and that 'the construct of race simplifies and exaggerates the role of heritable factors' (2003: 25; see also Sankar, 2006). In short, Cooper argues when race is defined biologically, it is too big a category—it includes too many people in any given racial category— to have any good predictive value. Other researchers warn that the association between race and health is leading to medical profiling in which certain racialized groups are regulated through medical surveillance (Leveque, 2008, forthcoming; Roberts, 2008; Schwartz, 2001).

At the University of Saskatchewan, Jennifer Poudrier (2007) shares this critique of research aimed at highlighting the association between race and

---

BOX 5.4 ❋ THE TUSKEGEE SYPHILIS STUDY

Steven Epstein (2007) observes that the infamous Tuskegee study of 'Untreated Syphilis in the Negro Male' was predicated on the hypothesis of biological differences between black and white men. In the early 1930s, United States government researchers, in collaboration with doctors at the Tuskegee Institute in Alabama, recruited almost 400 black men known to have syphilis and tracked the progression of the disease over four decades. The physicians observed these men endure symptoms including skin ulcers, blindness, motor coordination deterioration, bone structure deterioration, and death. Even when new research showed syphilis could be treated effectively with penicillin, this treatment was not given to the men.

Indeed, the research subjects were told they were receiving medical care for 'bad blood' and were not aware they were involved in a medical experiment. The men were actively discouraged from seeking medical assistance outside of the study. These men underwent invasive and painful tests including spinal taps.

Epstein draws attention to this study to demonstrate that it was predicated on a logic of racial difference. 'From the vantage point of the present era,' writes Epstein, 'in which it is frequently claimed that racial minorities will benefit from research that does not presume that whites and people of color are medically equivalent, the episode serves as yet another troublesome reminder that medical research premised on racial differences can sometimes serve stigmatizing and dangerous ends' (ibid., 43). Moreover, given that the Tuskegee study was not the first or last study of its kind, blacks in America amongst other racialized groups report high levels of mistrust of medical research, and are well known to be difficult to recruit as subjects in medical research. HBO recently produced a film—*Miss Ever's Boys*— about the Tuskegee study.

health. Analyzing the ways in which the thrifty gene theory—a theory that indigenous people are genetically predisposed to a certain form of diabetes because of a significant change from feast-and-famine to (essentially) feast-always conditions—reinforces established problematic binaries between 'civilized/primitive, Aboriginal/non-Aboriginal and science/culture'. Instead, Poudrier seeks to understand indigenous health from a 'decolonizing per-spective' (2007: 239, 253). By this, Poudrier means that scientific research on genes, race, and health must not simply take indigenous history and know-ledge into account: rather, it must begin from the perspective of indigenous peoples themselves.

Poudrier and other researchers are concerned about the ways in which ge-netics research might replay *eugenics* agendas. Eugenics is the theory that a better human species will be produced by allowing only some humans to sex-ually reproduce. Most scientists and politicians discredit eugenics but there are fears that continued scientific interest in human genetics might be used to support eugenicist political agendas. The Human Genome Diversity Project (HGDP) is a good example.

The Human Genome Diversity Project

The Human Genome Diversity Project (HGDP) attempted to map the ge-netic diversity within the global human population. This difference is less than one per cent (see Box 5.3). Stanford University's Morrison Institute, which started the HGDP, emphasized the potential benefits of this genetic mapping, including better understanding and surveillance of diseases associated with particular racial groups, as well as the origin of individual racial groups (i.e., how they evolved from a common ancestor). Ultimately, the HGDP failed when the desire of geneticists to sample the human genome clashed with indigenous groups (Reardon, 2005).

Critics of the HGDP—and there are many—argue the HGDP raised serious ethical concerns. A major criticism was that it would lead to racism because governments (and other groups) would be able to define and then target par-ticular racial groups through genetics. Reminiscent of the kinds of racial profil-ing that occurred in Nazi Germany, certain groups of people could be defined through genetic testing and then conferred (if they belong to the right genetic group) or denied (if they belong to the wrong genetic group) certain rights.

A second criticism is that the HGDP is itself infused with racist presupposi-tions. This criticism is based on the fact that the HGDP developers describe the project in similar ways as settlers described the peoples they colonized—the 'vanishing Indian' for instance. This issue is closely tied to that of informed consent. Critics argue that the HGDP did not adequately explain to the various indigenous populations whose DNA it has obtained what exactly their DNA is being used for, nor what it might be used for in the future (the information is kept indefinitely in DNA databases). There are significant cross-cultural issues here, including language barriers and cultural notions about what DNA is and how it relates to an individual's, as well as a culture's, identity. Poverty may play a significant role, as racialized groups who have historically been disadvantaged are more likely to be impoverished and therefore consent to giving a DNA sample in exchange for money in order to buy essentials like food for their family.

And there is no guarantee, critics further argue, that corporations would not patent the DNA information and use it for commercial profit. This has already happened, as pharmaceutical companies have patented drugs marketed to particular racialized groups (for instance, the company NitroMed markets the drug BiDil for treatment of heart failure in African Americans). It is not too far a stretch, some believe, to imagine the information might one day be used to develop biological weapons to target particular groups of people based on unique DNA. There is at least one known example of a scientist proposing the use of DNA material collected from indigenous populations to make forensic identifications (Kidd et al., 2006). Indigenous populations may well have no idea that their DNA samples may be used for these purposes.

For these reasons, sociologists link the HGDP to biopiracy—usurping indigenous knowledge for financial gain (Subbiah, 2004; Greene, 2004). And even in cases where western racialized people are availing themselves of genetic testing, as in the case of the individuals Nelson studied who underwent genetic testing to better determine their own ancestry, 'it does nothing to challenge the classificatory logic of human types that compounds, rather than challenges, social inequality' (2008: 776).

Indigenous Peoples and Environmental Issues

Concern for the environment, and the specific impact of human attitudes and behaviours towards other organisms and Earth, features in every culture throughout human existence. In the late 1940s and into the 1950s, for example, deep concern with the invention of the atom bomb and its effect on the future of humans, other animals and the planet, occupied the conscience of at least one generation (see Chapter 1). Albert Einstein, who in 1939 warned US President Franklin D. Roosevelt that the Germans were attempting to construct an atomic bomb and urged the US to begin its own bomb development, later devoted himself to cautioning people about the devastating effects of this particular turn in the arms race. Today, attention has shifted to global warming, and animal and plant species extinction.

The particular shape that environmental concern takes in any given generation is intimately dependent upon social factors such as culture, politics, and economics. As Chapter 1 discussed, western science and technology has always borrowed, taken, imitated, and otherwise incorporated knowledge about nature from indigenous and non-western cultures. The pressing issues of global warming, species extinction, sustainable agricultural practices, and environmental degradation in general may be making this incorporation explicit. Canadian government officials, researchers, and members of the public have begun to call for a more open exploration of traditional indigenous knowledge about the environment. Some argue the incorporation of indigenous knowledge is little more than paying lip service to First Nations, Inuit, and other indigenous peoples' expertise (Agyeman, Cole, and Haluza-Delay, 2009), while arguments that indigenous people do not enjoy a historically separate or special kind of relationship with the environment have met with critiques.

In *Native Science: Natural Laws of Interdependence* (2000), Gregory Cajete contrasts western and indigenous ways of understanding the environment. Arguing that western science is mechanistic—it seeks to determine causal laws underwriting the cosmos—Cajete argues indigenous knowledge allows a much more holistic approach to understanding the relationship between humans and our environment. Indigenous knowledge, writes Cajete, 'attempt[s] to gain comprehensive understanding rather than piecemeal analysis; a pursuit of wisdom rather than the accumulation of data . . . ' (2000: 61).

Native science understands humans to be part of the environment rather than somehow standing outside or apart from the world. Put another way, all living organisms, and Earth itself, are codependent upon all other living and nonliving organisms. So far, indigenous science closely resembles that of Darwinian evolutionary theory, Gaia theory, and other theories that emphasize the co-evolution of Earth. It departs company with evolutionary theory and takes on a more familiar western hue by identifying humans as somehow more responsible for (and thus more in control of) the fate of Earth and other species. 'The new cosmology' writes Cajete 'must reflect realization that the fate of the Earth is now intimately intertwined with the fate of the human species. We are the universe and the universe is us. . . . *We are the Earth being conscious of itself*' (2000: 61, emphasis added). The supposition that humans are somehow separate from any other species—that humans are Earth being conscious of itself—and that we control the destiny of Earth is controversial. First of all, it runs counter to Darwinian evolutionary theory (see Gould, 2000; Hird, 2009). This theory also seems to inadvertently reinforce Kant's separation of humans from the universe, insofar as humans seem to be imbued with a distinct and special power to inhabit and speak for Earth.

Indigenous science parts company with Darwinian evolutionary theory, and western science generally, by adopting mysticism or spiritualism as a way of knowing. Mysticism is a way of becoming consciously aware of reality (Earth, nature, cosmos) through spiritual intuition. The universe, for Cajete, is in 'dynamic multidimensional harmony' (ibid., 64). It is common, within this philosophy, to attribute spirits to all living organisms. Indigenous peoples enjoy a unique relationship with the environment because, the theory argues, they are more directly connected with Earth (than other human beings are) through their spiritual relationship with it. This leads to a turning away from the 'all-consuming materialism' of western societies, towards 'taking responsibility for our relationship to the Earth' (ibid., 61). The supposition is that by adopting indigenous ways of knowing and living within the environment, humans can create a much more sustainable future.

Some researchers focus on the knowledge indigenous people have accumulated about sustainable living, without relying on knowing through spiritualism, mysticism, or religion. These researchers point to the ways in which indigenous peoples have learned, through generations of trial and error, to survive in specific environmental contexts (Flannery, 1994; Pyne, 2001; Raffles, 2002; Trafzer, Gilbert, and Madrigal, 2008; Vale, 2002). For instance, Aboriginal people are known to have worked with, rather than fought against, the dry climate in Australia, strategically setting and lighting fires in order to steer fires in particular ways, improve the land's condition, and maintain passages through the land for humans to traverse (Clark, 2008; Langton, 1999; Yibarbuk, 1998). These fires were also strategically lit to better capture and kill animals for food, clothing, and implements (Franklin, 2008).

This is knowledge accumulated by people who have lived within particular geographical regions over many generations. The knowledge, derived from trial-and-error, has at times been very different from the knowledge of how to live in other climates and geographies (for instance, Europe) that colonialists brought to the New World. 'Experience', writes Clark, 'counts. . .and as a rule the longer a people or a species or a mixed community has inhabited a region, the greater the range of variation they will be conversant with, the more attuned they will be to its characteristic shifts, peaks and troughs' (2008: 5). Mike Davis (2001) details the dire consequences when indigenous and colonial ways of coping with the agricultural environment in India clashed during the Victorian era (1837–1901). 'When one group,' notes Clark, 'has the power to impose inappropriate responses to environmental stress and extremity on another, these repercussions can be catastrophic' (2008: 2).

Focusing on different human experiences with the environment, and the ways in which all species survive through adapting to the viscidities of the environment is one of the main tenets of Darwin's theory of evolution. It provides a way of integrating the histories of human species with those of other species, geographies, climates, and evolutionary history without making claims about indigenous primordiality or spiritualism (Griffiths, 2000). It is a way of escaping representations of indigenous peoples as harmoniously living with nature (Massey, 2005). It also provides a way of keeping in mind that trial-and-error can be very costly when it goes wrong (as must have been the case at times) and of the 'debt—immeasurable and unredeemable—that

is owed to those unnamed others whose trials and labors ultimately . . . '
enabled current species (including humans) survival (Clark, 2008: 6). It is
also to acknowledge there are environmental events that no peoples—indigenous and settlers included—can either predict or adapt to, particular
Earthly events, such as the unfathomably devastating impact of meteorite
hits, which no amount of improvisation and adaptation can prepare us for
(Bryson, 2005).

(Dis)Ability Science and Technology Studies

Before typing this sentence, I interfaced with numerous technologies: alarm
clocks, toilet, eyeglasses, computer, iPod, gas heating, toaster, iron, bicycle,
and so on. The activities I complete each day involve and reveal my depend-
ency on technology.

The philosopher Martin Heidegger famously argued we only notice technology when it breaks down. And for the most part, we each get through our
days using myriad technologies that we rarely notice or, indeed, understand.
Technology is typically defined as human-made phenomena that 'direct human action, embodiment and thought in certain ways' (Lupton and Seymour,
2000: 1852). Since the adaptation of objects into tools, being human has
become dependent upon, to greater or lesser degrees, myriad technological
innovations. While my generation—old as we feel, and as old as we look to
you—marvels at the iPhone, my 80-plus-year-old father talks about the staggering technological advancement that the hoof pick (an implement used to
remove stones from horse hoofs) meant for his generation.

A perusal of the journal *Technology and Disability* reveals a largely un-
familiar set of technologies: Lightwriter, Polycom, lateral key grip, DISLIB
language guiding system, Robot-mediated Active Rehabilitation (ACRE2),
wheelchair simulators, electronic stance, and swing phase controlled knee
joint. These technologies and the interfaces they invite are the domain of
disability studies. The field of disability studies is richly diverse (see Davis,
2006; Thomas, 2007). What interests us here is how we might better understand bodies—'normal' and 'disabled'—as they are constructed by and
mediated through science. The introduction and widespread use of personal
computers, the Internet, virtual reality games, and so on raise new questions
about how we experience our bodies. Lupton and Seymour (2000) make the
point that analyses of body–machine interfaces assume bodies are free from
illness and/or disability (see also Mort, Finch, and May, 2009). Impaired
bodies, it may be argued, offer particular insights into science, technology,
and power.

A body using any form of technology is essentially adopting a prosthesis
to enhance its capacity: my eyeglasses and my father's plastic knee for instance. The Internet and the World Wide Web are prosthetics par excellence

because they 'expand our sphere of action, our instrumentality and our ability to influence, as well as our experience and ways of experiencing' (Moser, 2000: 215). James Cameron's film *Avatar* (2009) is an extension of this idea.

However, 'the severely damaged body, the body that is culturally designated as "disabled" compared with other bodies designated "normal", remains subject to a high level of stigmatization and marginalization' (Lupton and Seymour, 2000: 1852; see also Davis, 1995; Thomson, 1997). In other words, the disabled body's relationship with technology is more visible, more present, leading to potentially different embodied experiences and societal expectations of bodies with disabilities. One consequence is that people with disabilities are expected to benefit much more from technology, to readily avail themselves of technological advancements, and to be grateful for the intervention. Technology, indeed, becomes a means through which people with disabilities are expected to overcome their impairments. Technology, in other words, normalizes the disabled body.

The rehabilitative perspective often adopts this kind of deficit model (Roulstone, 1998). People with disabilities, on the other hand, tend to report a much more complex, nuanced, and ambivalent approach to technology. The diversity of responses is in part due to the fact that disability is a catch-all term, encompassing a huge swathe of lived experiences with impairment. These experiences also depend upon how various disabilities intersect with other important markers in society, such as social class, race, ethnicity, age, sexuality, and gender (O'Neill and Hird, 2001; Acton and Hird, 2004; Fuglsang, 2005; Moser, 2006). Time—how long an individual and the society in which they belong interfaces with a particular technology—is also an important factor (Winance, 2006). Also, the importance of history cannot be over-emphasized. Chris Degeling's (2009) fascinating study of the transformation of human and animal bone fracture care—for humans through economic rationalization, and for animals through the sentimentalization of particular kinds of animals into pets—reminds us that complex historical, economic, cultural, institutional, and social factors work together to produce particular societal values (and hence scientific interest) around technologies.

The diversity of responses is also reflected in the fact that technology encompasses a broad range of objects and relationships. For instance, eyeglasses and diagnostic genetic testing are both technologies but they differ considerably in their bioethical implications and experience of use (Hess, Preloran, and Browner, 2009).

Certainly, for people with disabilities, technology can offer ways of better navigating daily struggles (Goggin and Newell, 2004). On the other hand, technology breaks down, new developments in technology need to be constantly learned (and with only a very small proportion of the population familiar with specialized technology there are few sources of assistance), and the exorbitant cost of specialized technology complicates the assumption that technology is always a win–win situation for people with disabilities (Moser, 2000). Technologies may also further isolate individuals by providing the illusion of independence (Lupton and Seymour, 2000).

Technology, in other words, might well further the exclusion and marginalization of people with disabilities by actually increasing their dependence on technology, and the people required to use technology. As Moser writes:

> There is a certain ambivalence between the desire to help disabled people and improve their lives on the one hand, and the fact that differences are defined as lacks or deviations that ought to be corrected on the other. It is therefore perhaps not so surprising that we find all normalization strategies come with strings attached, the most important of which being the notion that such a thing as an ideal that we can call normal actually exists—the idea of a normal person, or the construction of a norm. (2000: 209)

Moreover, some people reject the use of technology because this technology reifies the societal assumption that particular individuals have disabilities. To wit, some members of the deaf community reject the use of cochlear implants because it represents deafness as an abnormality in need of correction (Davis, 1995; Yardley, 1997). From this perspective, technology reinforces the already pervasive societal assumption that disability is an individual (correctable) condition, while simultaneously drawing attention away from the socio-political dimensions of disability (Oliver, 1990). This discourse 'orders reality' as Moser puts it: 'first you are marginalized and excluded, so that in the next turn you have to be included and rehabilitated. At the same time you are doomed to fail . . . you will constantly be countered by processes that continue to produce inequality and exclude you—*ad infinitum*' (2000: 210).

Summary

Political economy, feminist, post-colonial, and disability science studies provide important insights into science and technology and/as power. These analyses remind us that science and technology are indelibly mixed with political economies. Political institutions, economic systems like neoliberalism, and cultural consumption and production patterns both determine and are determined by research and developments in science and technology. Science as power reminds us that we need to pay close attention to the bread-and-butter of science, including what and how science classifies entities, what and how research agendas are generated, who gets to do certain kinds of research, what research is funded, and how technologies are selectively developed and utilized within local environments (schools, for instance) and globally (from developed to developing countries, for example).

As well, debates within these fields generate innovative, timely, and complex questions about how to best focus sociological exploration and critique. Within feminist science studies, for instance, opinions differ as to whether to include (and acknowledge full rights to) nonhuman entities. Some feminist theorists worry the inclusion of more and more diverse nonhuman entities dilutes and may eventually entirely erase fundamental still-unresolved myriad issues facing women and girls in relation to science and technology, such as the use of women's bodies in scientific research and the uneven distribution of women in science disciplines. There is by no means a consensus within post-colonial science studies that race should become re-associated with biology, and genetics in particular. And (dis)ability science studies scholars are divided as to whether discussions of innovations in human cyborg technologies will aid people with disabilities through increased mobility and access to community resources or will further efface their lived experiences and identities.

Key Terms

Political economy This is a term used to describe the analysis of a state or group's process of economic production, including buying, selling, laws, customs, and government.

Colonialism This term refers to the expansion and maintenance of territories (including the peoples found within any given territory) by people from a different territory.

Capitalism This is an economic system whereby the means of production is privately owned for profit.

Human Genome Diversity Project Developed by the Morrison Institute in the United States, the HGDP multi-member scientific team sought to derive the genetic sequencing of distinct races and ethnicities.

Indigenous This term refers to a defining characteristic of peoples claiming affinity to a particular territory based on longest-established occupation of that territory before settlement by other peoples (through processes such as colonialism).

Discourse This term defines a system of signs or sequence of relations between objects often described in terms of written and oral communication.

Critical Thinking Questions

1. What does it mean to engage with scientific knowledge? Can sociologists do this without engaging in power relations?

2. What are the advantages and disadvantages of endorsing race-specific and sex-specific health research? For instance, should certain drugs be targeted to people belonging to certain races and/or women?

3. Should indigenous elders be allowed to sell their genomic information on behalf of the people they represent?

4. Why do people with disabilities have an ambivalent relationship with science and technology?

5. How might sociologists engage with scientists to better understand disability?

Suggested Readings

J. Agyeman, P. Cole, and R. Haluza-Delay, eds, *Speaking for Ourselves: Environmental Justice in Canada.* Vancouver: University of British Columbia Press, 2009. This Canadian collection provides timely and thought-provoking analyses of environmental social justice issues facing current and future generations of Canadians.

D. Haraway, 'Situated Knowledges: The Science Question in Feminism and the Privilege of Partial Perspective', in *The Science Studies Reader.* New York: Routledge, 1988, pp. 172–88. This well-known article argues knowledge is situated within particular contexts of time and space rather than objective in any kind of rarified way. It is a must-read within science and technology studies.

V. Satzewich and N. Liodakis, *'Race' and Ethnicity in Canada: A Critical Introduction,* 2nd edn. Toronto: Oxford University Press, 2010. This is a useful introduction to race and ethnicity concepts, theories, and methods, with an emphasis on the Canadian context.

Kimberly TallBear, 'Native-American-DNA.coms: In Search of Native American Race and Tribe', in Barbara Koenig, Sandra Soo-Jin Lee, and Sarah Richardson, eds, *Revisiting Race in a Genomic Age.* Piscataway, NJ: Rutgers University Press, 2008, pp. 235–52. This article asks difficult and timely questions concerned with indigenous sovereignty in the context of contemporary biotechnology.

Websites and Films

PBS NOVA DVD, *Judgment Day: Intelligent Design on Trial*
www.pbs.org/wgbh/nova/evolution/intelligent-design-trial.html
This is an excellent summary and analysis of the history of the Intelligent Design debate within North America. It also distills the salient arguments used on both sides of the debate.

The Human Genome Controversy

www.wired.com/wired/archive/5.07/updata.html

This website provides a useful summary of the controversy surrounding the Human Genome Project. It provides a history of project and how it courted controversy throughout its tenure.

Secret of Photo 51

This PBS NOVA film provides a detailed account of Rosalind Franklin's life as a scientist within a male-dominated university system, and her groundbreaking research that led to the production of the DNA double helix.

6

Values, Trust, and Public Engagement with Science and Technology

Learning Objectives

In this chapter we learn:
- Science and technology both reflect and influence societal values;
- Science and technology are both authoritative and contested;
- Society is shifting to a new social contract between science and society that focuses on greater public engagement with science;
- Science and technology actively manage this social contract through social relations.

Good Science, Bad Science

During the late 1930s and early 1940s, at the height of the Second World War, the Nazi party authorized doctors and scientists to experiment on concentration camp prisoners (Nicosia and Huener, 2002; Proctor, 1990; Weindling, 2006). Some of these experiments were ostensibly designed to gain new knowledge about the conditions faced by German soldiers. Most, if not all, of these were called terminal experiments. Unlike medical research, the test subjects were not meant to survive the research (Weindling, 2006).

The 'experiments' included freezing: concentration camp inmates were left immersed in tanks of ice water until they froze to death in order to see how long German pilots might survive if they were shot down over the North Sea (Pellegrino and Baumslag, 2005). Other inmates were exposed to decompression chambers to simulate high altitude conditions, and their brains were dissected while they were still alive. Other experiments used chemicals, poisons, wounds, artificial insemination, sterilization, and surgery.

In my early days as a sociologist, I attended a lecture given to first-year sociology students at Auckland University (New Zealand) by one of my colleagues, Ian Carter. I attended the lecture because of its reputation. Carter, a scholar interested in the sociology of transportation, and train routes more specifically, delivered a memorable lecture on the technological development of trains and how their deployment across Europe (and later North America) enabled profound economic, political, and social change. As he was describing the train system that brought victims to their deaths at Auschwitz and other concentration camps, Carter drew attention to the Nazi experiments. Carter acknowledged that he stood before the class having recently survived triple bypass surgery. Some of the surgical techniques that enabled Carter to survive coronary heart disease were, in part, developed at Auschwitz (Nicosia and Huener, 2002).

Carter brings to our attention the shifting relationship between science, technology, and values in society. Some people maintain that what the Nazis did was not science at all: it was torture. The experts who investigated the Nazi research for the Nuremberg Doctors' Trial noted the perversion of scientific terminology: 'control subjects' endured the most suffering and died, 'sample size' referred to hundreds of Jewish prisoners, 'significant findings' meant the level of suffering, and 'response rate' referred to the level of torture (Cohen, 2010: 19).

Important questions are raised about the scientific validity and reliability (see Chapter 3) of this data. For instance, the experiments were conducted on people who were starving and, as a result, likely had other health issues (such as compromised immune systems, parasitic infections, and severe psychological trauma). Can any results, then, be generalized from this very specific population to people as a whole? More broadly, since the experiments were conducted unethically, can the results be reliable (Moe, 1984)?

Some researchers argue research that serves an explicit political agenda is necessarily unsound and therefore constitutes unethical or bad science (Buchanan, in Moe, 1984). For instance, an analysis of the hypothermia data generated by Doctor Sigmund Rascher at Dachau found inconsistencies (related to how long it took to kill frozen prisoners) in his laboratory notes. Historians suggest these inconsistencies occurred because Rascher was ordered by Himmler to produce hypothermia results quickly. (When Himmler discovered Rascher's manipulation of the data, he had Rascher and his wife killed; see Cohen, 2010).

On the other hand, and as Carter alluded to in his lecture, what if findings derived from this horrible period of human history can be used to benefit society by saving lives today? A number of contemporary scientists, such as John Hayward, an Emeritus professor at the University of Victoria in British Columbia, have used Nazi data to further their own research. Hayward studies hypothermia with a view to developing better cold-water survival suits for people on fishing boats on the Canadian coastline. He has said of the hypothermia experiments performed at Dachau, 'I don't want to have to use this data, but there is no other and will be no other in an ethical world. I've rationalized it a little bit. But to not use it would be equally bad. I'm trying to make something constructive out of it' (Moe, 1984: 272). For instance, the Nazi research discovered the technique of Rapid Active Re-warming for frozen people, a finding that contradicted the prevailing traditional and not very successful method of slow passive re-warming (Cohen, 2010).

BOX 6.1 ❈ THE NAZI WAR ON CANCER

Robert Proctor's book, *The Nazi War on Cancer* (2000) provides a useful example of the complicated and ambivalent relationship between science, technology, and values. Proctor's book documents the keen interest that Hitler and his colleagues had in creating a superior Aryan race. Towards this goal, Hitler urged German scientists to conduct research on how to prevent and treat cancer. The Nazi party strongly encouraged Germans to give up smoking and eat a vegetarian diet focused on locally produced, organic food. The roots of the contemporary organic food movement are contained in the Nazi platform on healthy eating. Hitler also ordered research to be undertaken on occupational health and safety. This research focused on occupational carcinogenesis, including radium, uranium, various dusts (such as arsenic), and asbestos. The Nazi party argued strongly for health and safety awareness and policies inside Germany's factories.

This research must be understood in the context of the overarching Nazi plan to exterminate millions of people in the name of racial purity and superiority. It draws attention to the complicated relationship between science, technology, and values. Contemporary western society maintains a rhetoric of the moral superiority of organic, locally grown food. Canada has numerous laws and policies related to health and safety in the workplace. Is it relevant that the seeds of this moral valuation, policies, and research on cancer and carginogenesis originated at least in part in the Nazi regime?

What happened at the Nazi concentration camps is a particularly graphic illustration of the co-production of science and technology, and societal values. As this book argues throughout, science and technology are social relations, and they necessarily engage with issues of trust, credibility, and power.

Science and Values

Modern science is the key representation of the Enlightenment ideal of a society based on rationality. As such, science explicitly values reliability, validity, precision, accuracy, testability, generality, concept simplicity, and hands-on learning (Allchin, 1998). Science explicitly *dis*values pseudoscience, fraud, and error. Therefore, science is not value-free.

American sociologist Robert K. Merton developed what became known as the Merton Thesis, and a theory about the norms of science whose acronym is known as CUDOS. The Merton Thesis (1973) advances a positivist view of science. It argues that the scientific revolution (see Chapter 1) occurred because of the development of better experimental techniques and because naturalists accumulated more observations about nature (Becker, 1992; Shapin, 1988). The Merton Thesis also argues the scientific revolution in seventeenth-century England was positively correlated with the high proportion of Protestant (and often Puritan) scientists who dominated the Royal Society. Protestantism, Merton argued, is positively correlated with the values of this new science, which became the foundation of modern science. These values included empiricism and rationalism in identifying the (God-given) order of the world. Protestantism provided a justification for carrying out scientific research because scientists charged themselves with better understanding God's plan for humanity and the world. Prior to the Enlightenment, curiosity was strongly associated with evil. In the Christian Bible, the Devil (disguised as a snake) encourages Eve's curiosity about the tree of knowledge, which was itself a symbol of humanity obtaining knowledge about the world. God's plan, according to Christianity, was not for humanity to know, and to enquire was to show conceit. Francis Bacon (see Chapter 2) strategically distinguished two kinds of knowledge: knowledge about God and knowledge about the world. Wanting to know about God, Bacon argued, showed conceit; but wanting to know about the world showed a healthy kind of curiosity. The Merton Thesis draws attention to a dramatic shift in the Christian view of the relationship between morality and a desire to understand the world.

According to Merton (1973), four ideals organize scientists' practice of science, identified by the acronym CUDOS. Communalism refers to the ideal that scientists exchange ownership of their scientific discoveries for recognition. Universalism refers to the ideal that claims about the world are evaluated using universal and impartial criteria, rather than criteria such as the wealth or status of the scientist. Disinterestedness refers to the ideal that scientists are esteemed for acting in selfless ways (for the common good of scientific knowledge). Finally, Organized Skepticism refers to the idea that the scientific community will rigorously test all claims to knowledge.

Scientists conduct their research according to these norms, Merton argues, in order to produce knowledge that society agrees is credible and trustworthy. Scientists, according to Merton, should not let personal or social attributes (ethnicity, culture, gender, religion, social class, and so on) interfere with either the research process or the conclusions. Scientific research must not be valued according to authority but rather agreed upon through the accountability of peer review. Science should involve methods that explicitly evaluate knowledge claims such as controlled observation, experiment, testability of predictions, repeatability, and statistical analysis (Allchin, 1998). These ideals and norms have clear origins in the Enlightenment. Science attempts to regulate its members through the internalization of these norms and reiterates these norms when it engages with members of the public in order to increase levels of trust and credibility.

Science and Trust

Canadians trust science to provide facts about our bodies, minds, lives, and world. Science tells us things about the food we eat, the air we breathe, the cars we drive, our life spans, relatives, feelings, animal pets, and our environment. So do we trust in science, in all circumstances, and at all times? Who and what do we trust when we trust science? Do we need to know how science works in order to trust it? Do we care how science works?

We know that the public's trust in, and appreciation of, science grew after the Second World War, and has continued apace—in a somewhat more critical vein—since (Lewenstein, 1992). A number of social scientists empirically (through surveys) track how different cultures, at different times, understand and react to science (see, for example, Bauer and Schoon, 1993; Durant, Evans, and Thomas, 1989; Luján and Todt, 2007; Miller, 1991). We know from surveys that western societies generally report high levels of trust for scientists and those who apply scientific knowledge—especially health care professionals (Miller, 2004). Science, these surveys show, is believed to be an objective and unbiased source of truth, and working toward a better future (longer lives, solutions to problems such as world hunger and disease, etc.). By and large Canadians trust in scientists' self-regulation and expertise.

In *A Social History of Truth: Civility and Science in Seventeenth-Century England* (1994), Steven Shapin observes that science and trust did not always go together. Indeed, scientists have only recently been characterized as worthy champions of valid and reliable truth. Recall from Chapter 1 that scientists did not exist at one time. Instead, a diverse group of people used a wide range of theories and methods to make claims about the natural world. This group, writes Shapin, included court magicians, faith healers, and the like. Only in the eighteenth and nineteenth centuries did the scientist as a particular occupation and expert identity emerge, and this emergence occurred in tandem with government institutions and imperial conquests. And ironically, argues Shapin, the very successes of science—for instance in weapons development and medicine—have provided the impetus and foundation for scientists, universities, and scientific societies to be co-opted by governments and business to pursue and legitimate political and economic agendas (see Chapter 5).

This is not to say people are always and uniformly enthusiastic about science. For instance, surveys show that most people still believe in God (see Chapter 1). Surveys also show that health research findings are not the sole, or even main, determinant of patients' beliefs or actions; patients and members of the broader public garner evidence from an increasingly diffuse range of sources including the Internet, magazines, friends, family, and acquaintances (Levin, 2008). Tracing British attitudes towards biological research, Bauer (2007) traces definite peaks and troughs from the 1940s to the 2000s. What Bauer calls 'watershed events'—genetically modified soya and Dolly the sheep—did not initiate skepticism towards science but rather catalyzed an already established trend in the 1990s towards increased skepticism (ibid., 40). They reflect a dual process of increasingly open systems of knowledge production and the growth of complexity and uncertainty in society (Nowotny, Scott, and Gibbons, 2001).

One important area to consider is the ways in which different media take up, translate, and otherwise mediate scientific knowledge for lay consumption. A number of studies (for example Pellechia, 1997; Görke and Ruhrmann, 2003) show media accounts of science omit both contextual and methodological information, while also emphasizing possible, potential, and significant long-term consequences of the research. Other studies suggest media sometimes downplay scientific uncertainty because they think this will undermine calls for collective action, for instance, around issues of global warming (Olausson, 2009).

Journalists covering science stories relate experiencing persistent tensions between the desire and need for media to sell stories; the desire of scientists to control what journalists write about them and their research; the injunction that journalists inform and not educate their consumers; and a keen awareness that people are asked to take up and interpret often complex and fragmented accounts of scientific research (Dunwoody, 1992; Saari, Gibson, and Osler, 1998). Sociologists are interested in how people interpret medical controversies through media accounts, how they negotiate conflicting medical information, and how exposure to medical uncertainty can actually help people better appreciate the complexities of the issues involved (Dixon et al., 2009).

Public Understanding of Science and Knowledge Translation

In 1985, the (British) Royal Society launched a public understanding of science program, whereby the public was to be informed of, and educated about, important scientific and technological developments in order that the public better support scientific innovations. Calls for increased public understanding of science are made on the basis that members of the public: (1) have little understanding of science, (2) need to be educated about science, and (3) will more strongly support science once they have been educated (Field and Powell, 2001; Wynne, 1992; Irwin and Wynne, 1996). This is known as the deficit model.

Research suggests people generally have a low level of scientific literacy (Miller, 1991, 2004). In response, western governments spend millions of dollars trying to increase scientific literacy amongst their citizens, using school curricula, television programs, and public lectures to get messages about the goals and content of science across (Popli, 1999). Governments do this because it is widely accepted that familiarity with scientific concepts and principles is a pre-requisite for effective democratic decision-making (Miller,

1991, 2004). To give one example, Sturgis, Cooper, and Fife-Schaw (2005) conducted a large empirical study of peoples' knowledge of biotechnology and found that scientific knowledge has a positive impact on individual and group attitudes towards genetic science.

The deficit model is associated with knowledge translation. Both are in turn associated with a top-down approach to knowledge according to which experts are assumed to have knowledge that they translate to various audiences (politicians, policy-makers, media, and the general public). The success of knowledge translation is measured by the extent to which these audiences faithfully take up the knowledge as the specialists intended.

Knowledge translation is a key axis within Canadian health research, as basic and clinical researchers strive for the most effective means to get their findings 'from bench to bed' (Armstrong et al., 2007; Davis et al., 2003, Genuis and Genuis, 2006; Tugwell et al., 2006). Yet as much as clinical researchers may want patients and members of the general public to follow recommendations emanating from clinical research, scientists are aware that knowledge does not necessarily, or often, circulate in this way. Patients and the public mediate the knowledge they receive from family doctors and other health care professionals. People use multiple sources of information, from the Internet, magazines, and television shows such as *The Dr. Oz Show, The Doctors,* and *Dr. Phil,* to information provided by friends and family.

People mediate information according to their level of education, social class, cultural background, gender, and other factors. As such, there is no single public; there are, instead, several publics. Other factors need to be considered too. Powell et al.'s (2007) research found emotions also play a part in mediating scientific knowledge, with worry and anger being associated with perceived uncertainty about the implications of scientific research for the individual. Perceived lack of knowledge and perceived likelihood of becoming ill are only weakly associated with perceived uncertainty, however. The researchers found various demographic variables (education, socio-economic status, and so on), as well as risk contexts, communication processes, perceived knowledge, emotions, and perceived uncertainties about risks all played a role in attitudes toward science.

People also mediate scientific information when they consider a particular perceived scientific uncertainty (for instance, genetically modified food) alongside other perceived scientific uncertainties (such as Creutzfeldt-Jakob—mad cow—disease). In this context, people will rank any given perceived risk as either low or high (Townsend, Clarke, and Travis, 2004). The deficit model, again, does not take this kind of relativization of risk into account (see also Spence and Townsend, 2006).

Language also mediates how scientific knowledge is understood. Lorraine Whitmarsh (2009), for example, conducted a qualitative study on people's understanding of, and attitudes towards, climate change and global warming.

Whitmarsh found that people were more likely to absolve themselves of causes, impacts, and responsibility for tackling environmental issues when they were termed under the umbrella 'climate change' rather than 'global warming'.

Scientific Controversies

The public's trust in science is mediated by scientific controversies. Scientific controversies refer to disagreements about the validity and reliability of a scientific claim. We know from Fleck and Kuhn (see Chapter 2) that science does not always or only work through the progressive addition of facts. And facts themselves are social productions. As such, controversies arise because science works within paradigms that involve conventions, norms, and values.

In the scholarly literature and in the media, scientific controversies tend to be of three different (although sometimes overlapping) sorts (Engelhardt and Caplan, 1987). First, controversies may be concerned with the validity of a particular scientific discovery or invention. The controversy over the supposed discovery of cold fusion is a good example. In 1989, Stanley Pons and Martin Fleischmann announced to the media that they had produced cold fusion in an apparatus that fit on a countertop. This was an amazing claim, since fusion is what makes the sun's energy and the hydrogen bomb. In the following months, chemists and physicists attempted to replicate the researchers' findings: some were able to replicate the findings, while others were not. Theoretical chemists and physicists also came forward with contradictory claims about the abstract feasibility of such an experiment. In the end, the weight of scientific opinion sided against cold fusion. Sociologists (for example, Collins and Pinch, 1993; Lewenstein, 1995) examined this controversy in terms of how controversies begin and end in science. Controversies begin, they argue, when scientists produce unexpected (and sometimes fantastic) results. Controversies end when the weight of scientific consensus either confirms or condemns the research.

A second type of scientific controversy concerns arguments over categorization or classification (recall the discussion of classification in Chapter 1). This type of controversy ranges from topics that seem rather unimportant and arbitrary to the public (the classification and then declassification of Pluto as a planet is a case in point), to classifications that involve humans. The history of science reveals racialized people, women, and other groups of people were categorized as inferior to Europeans (and European scientists) (see Gould, 1996).

From a sociological perspective, one of the interesting characteristics of these controversies is their retrospective aspect. That is, at the time, these studies tended *not* to court a lot of controversy because they conformed to the morals and political agendas of the age. Only now do these scientific studies, and their methods and findings, court controversy—especially if they are referred to in contemporary arguments about racial or gender inferiority. This begs the question, of course, as to what contemporary scientific research will be considered controversial in the future.

The final type of scientific controversy concerns scientific misconduct. These controversies occur when scientists are judged to have misled the scientific community and the public about crucial aspects of their research. A recent example concerns the controversy generated by Woo Suk Hwang, who in 2004 published a much-publicized article in the respected journal *Science*. The article detailed a study he and his South Korean team conducted to clone the first human embryonic stem cell line. It transpired, after months of investigation by the media and by Hwang's university, that he had fabricated the results and used women's embryos without their permission. A number of sociologists analyzed this case in terms of how the controversy was eventually

---

BOX 6.2 ☀ CLIMATEGATE

In 2009, thousands of emails and other documents from the University of East Anglia's Climate Research Unit (CRU) were released over the Internet. The resulting scientific controversy is known as the Climate Research Unit email controversy, or 'Climategate' (after the political Watergate controversy in the 1970s). The emails revealed scientific misconduct, according to some critics. The controversy—and it has not to date been resolved—focuses on a fraction of the emails that discuss how to challenge climate change skeptics.

One email refers to a 'trick' used by Michael Mann (a climatologist at Pennsylvania State University and the creator of what is known as the 'hockey stick graph' of temperature trends) in his graph to 'hide the decline' of temperatures in the 1950s, when they are known to have increased (Pearce, 2010). These phrases were taken by critics to mean that climatologists faked data in order to support a political agenda. They also charged the climatologists with collusion, withholding scientific information, interfering with the peer review process to prevent scientific papers that disagreed with their findings from being published, and preventing data from being revealed under the Freedom of Information Act (Hickman, 2009; Revkin, 2009; Webster, 2010; Johnson, 2009; Mooreg, 2009; BBC News, 2009, 2010).

Scientific experts concluded that '[t]he so-called "trick" was nothing more than a statistical method used to bring two or more different kinds of data sets together in a legitimate fashion by a technique that has been reviewed by a broad array of peers in the field' (Foley et al., 2010). This controversy has again raised questions about the integrity of science, and politicians have called for public inquiries to re-establish scientific credibility.

settled—Hwang was convicted of fraud and embezzlement, fired from his university, and given a two-year suspended prison sentence—and how scientists, the Korean government, and the media separated Hwang's research from stem cell research in general, sustaining the promise of stem cell research to cure diseases (Kitzinger, 2008; see also Moore and Stilgoe, 2009).

Mike Schäfer (2009) argues the media tends to characterize scientific research as more or less controversial, depending on the topic. Stem cell and human genome research, for instance, is described and engaged with in more controversial terms than neutrino research. In a related vein, a number of sociologists are interested in the ways in which media employ science fiction as metaphoric illustration of contemporary biotechnological issues. Using the example of two prominent cases in the UK in which two families tried to use stem cells from a 'perfectly matched sibling' to treat their diseased children, Petersen, Anderson, and Allan (2005) showed how science fiction and fact weave together in media accounts to support particular public deliberations about scientific controversies and their impact on society, and that the imagery of science fiction explains why people have such strong responses to biotechnology and its imagined consequences for society (see also Ferreira, 2004). Stern (2006) and Scott's (forthcoming) research into both the Body Worlds exhibition of plastinated human bodies and the Visible Human Project shows that lay people themselves produce strong emotional reactions and feelings of ambivalence towards the ethics of particular aspects of science, such as autopsy and techniques for dissecting bodies.

Another interesting issue to consider is the degree to which scientists themselves trust one another, as well as their own disciplines. Scientists' own relationship with the ideals of science—truth, objectivity, dispassion, rigour, and social progress—is complex and necessarily interpolated within wider normative structures (see Chapters 1 and 5). A number of examples point to the fact that scientists in various contexts and at different times trust and don't trust each other (see Engelhardt and Caplan, 1987). These examples also provide insight into the ways in which scientists interact with each other's work, and how science monitors itself.

Knowledge Mobilization and Citizen Science

Knowledge controversies have helped shift science's communication with publics from knowledge translation to knowledge mobilization. As Chapter 5 examined, science and technology have traditionally been the purview of select members of society including the scientific community, industry, and government. Members of the general public have traditionally been kept out of the decision-making loop. The presupposition that there is a single homogenous public that can be made more knowledgeable about science for the dual purpose of increasing public support for science and increasing people's conformity to particular attitudes and behaviours has courted much criticism. Not only do sociologists critique the presupposition that increasing this kind of public conformity is necessarily in the public's best interest, but research shows knowledge translation is not particularly effective.

Sociologists produce important analyses of the limitations of top-down knowledge translation models. Brian Wynne (1996), for example, analyzed the clash between scientists, government officials, and farmers in the UK over the effects of radiation fallout from the Chernobyl nuclear disaster in 1986. Experts, backed up by politicians, told sheep farmers in Cumbria (an area in northern England) to restrict the areas in which their sheep grazed in response to radiation fallout from Chernobyl for a particular duration. The farmers, however, were less concerned about Chernobyl (in Ukraine), and more concerned about Sellafield, a nuclear processing plant on their door-step. The farmers had local expertise about the effects of this local radiation on the land and their sheep that the experts and government ignored. Wynne found this was not a misunderstanding: the scientists assumed themselves to be the only source of expertise (and the politicians agreed), and the farmers distrusted the scientists and politicians as a result. Moreover, it turned out that the experts were wrong, and their miscalculations cost the farmers hundreds of thousands of dollars in lost revenue (see also Collins and Evans, 2002).

Knowledge mobilization is emerging as a response to the need to better engage with people about science and technology. It refers to the idea that there needs to be a more equalized communication between publics and those who have knowledge about, and power over, the particular issues at stake, including scientists, policy-makers, and politicians (Joly and Kaufman, 2008). Knowledge, in other words, needs to be mobilized within and between concerned groups of people. At its core, knowledge mobilization seeks to explicitly gather together the specialized knowledge of lay people (such as the farmers in the case above), scientists, and so on, as well as the concerns of community groups, policy-makers, and politicians, in order to most effectively address public issues of concern. As such, knowledge mobilization is strongly associated with citizen science.

Various centres, grassroots organizations, and academic research projects provide an important space for citizen participation in scientific research and its applications (Nowotny, 1993). These efforts are founded upon the recognition that knowledge is not only socially constructed, but has serious social repercussions for particular groups and society in general. In some cases, panels of lay people are convened to provide a public forum in which people question scientists and other experts in order to form their own judgments about policy directions. The Canadian Institutes of Health Research, for instance, fund a Café Scientifique program, where health care practitioners gather with members of the public to discuss a particular health topic. In other instances, non-expert consumers work directly to design alternative technologies better suited to their particular needs. In still other instances, university science and technology studies courses are based around participatory activities as an integral part of their learning programs.

In the Netherlands, a number of university-based community research centres produce studies each year for community groups, trade unions, and public-interest organizations (Sclove, 1996, 2001). These centres are found throughout Europe and North America. Denmark conducts directed consensus conferences in which lay people participate in intensive workshops to become informed about a particular topic, after which they discuss their ideas and judgments at national press conferences attended by members of Parliament (Sclove, 1996; Einsiedel, Jelsøe, and Breck, 2001).

At Oxford University's School of Geography and the Environment, Sarah Whatmore (2009) and her colleagues convene competency groups involving community groups, individual concerned citizens, scientists, social policy analysts, and local politicians. These stakeholders meet on a regular basis and with an agenda agreed upon by everyone, for extended periods (months to years) to work out the best way to solve a particular environmental problem (such as flooding). Whatmore and her colleagues map the language stakeholders use in order to trace the 'partisanship' of scientific knowledge claims (ibid., 587); the entanglements of scientific, legal, moral, economic, and social claims; and feasible ways of intervening in knowledge controversies (see also Farkas, 1999; Jasanoff, 2005; Murphy, 2006; Warner, 2008). Inger Lassen (2008) provides another model for focus group discussions about genetically modified food that include members of the general public and biotechnology scientists.

David Hess developed the term 'undone science' to refer to the absences in scientific research that social movement and civil society groups find when they try to find research on particular topics of concern. When community groups challenge a particular existing or proposed entity—nuclear energy, landfills, or industrial turbines, for instance—they often find a lack of research they may harness in favour of their claims. This is not surprising given the fact that industry, government, and the military almost always have access to much greater financial and other resources to support scientific research that will favour their own agendas (Blume, 1974; Cummings, 1984). Hess notes this selective funding has led to the uneven development of scientific research in favour of conducting research of benefit to elites but not necessarily the broader society. Political and economic elites do not enjoy a complete stranglehold on scientific research, however, and there are examples of civil society groups affecting research agendas. This was the case, for instance, when AIDS patients were able to secure a place on the agenda-setting panels of the National Institutes of Health in the United States (Epstein, 1996). Some civil society groups are large and wealthy enough to fund scientific research themselves, and gather together the expertise they need to advance their own research-backed agendas. It is important to point out, as Hess (2009) does, that there is no guarantee that civil society research represents the best interests of society—it is quite possible to serve much narrower interests (Hess uses the example of spiritualist mind–body therapies).

Moreover, even when civil society groups are able to get undone science done, there is no guarantee the research will not be stigmatized or otherwise undermined by other (elite) interest groups (Frickel et al., 2010). Sociologists analyze the widespread and successful practice of 'manufacturing uncertainty' used by industry to cast public doubt on scientific research findings that show negative health consequences of that industry's product. The most notorious example here is the tobacco industry's long-standing efforts to cast doubt on scientific research showing the association between smoking and cancer. (See French and Hird, 2008, for an analysis of how the gambling industry in Canada has manufactured uncertainty about the negative effects of gambling). In response to a report of the American Cancer Society, the tobacco industry emphasized three basic points: 'That cause-and-effect relationships have not been established in any way; that statistical data do not provide the answers; and that much more research is needed' (Confidential Public Relations Report, cited in Michaels and Monforton, 2005: S40). The tobacco industry even went as far as to establish a journal in the 1960s, called *Tobacco and Health Research*, the goal of which was to cast doubts on the 'cause-effect theory of disease and smoking' (*Tobacco and Health Research*, cited in Michaels and Monforton, 2005: S40). The lead industry, the chemical industry, and the asbestos industry all followed suit. Building on a strategy developed by the American public relations firm, Hill & Knowlton, the general approach was to dispute claims on the basis of the non-representativeness of studies in the human population and the non-relevance of studies in animal populations. These declarations were accompanied by the assertion that more research was needed (ibid., S41).

This amounts to an industrial-led (and funded) attempt to brand science, which acknowledges uncertainty as a fundamental element of scientific research (see Chapters 3 and 4), as junk science. For instance, Philip Morris (a tobacco corporation) secretly funded an organization called the Advancement for Sound Science Coalition (ibid., S43). Michaels and Monforton write:

> It is difficult to find a meaningful definition of the term 'junk science.' Peter Huber, who is often credited with coining the term, offers a broad-ranging 'I know it when I see it' description rather than definition: 'Junk science is the mirror image of real science, with much of the same form but none of the substance . . . It is a hodgepodge of biased data, spurious inference, and logical legerdemain. . . . It is a catalog of every conceivable kind of error: data dredging, wishful thinking, truculent dogmatism, and, now and again, outright fraud. (ibid., S43)

It is worth also noting that the term junk science is not used in the context of scientific debate, but is instead deployed with maximum efficacy in broader public forums—its rhetorical edge could be too easily turned back upon its wielders in a context where they too were accountable to scientific standards (ibid., S43). While the meaning of junk science disappears when probed (its articulators have never offered a method for distinguishing real science from junk science), and while its articulators are often in the employ of industry trying to avoid regulation and litigation, the increasing use of the term in policy circles must also be seen as related to 'the very nature of scientific evidence dealing with human behavior' (ibid., S43). Manufactured uncertainty raises key questions about how we may discern knowledge that is valid and reliable, and how we, indeed, may discern genuine controversies in sciences from manufactured ones.

Gross (2007, 2009) raises an additional consideration, that of negative knowledge—knowledge deemed dangerous or not worth pursuing. Decisions about what knowledge is negative knowledge are typically made by elites. For instance, legislators and federal agencies will set limits on the scope of research programs such as cloning, stem cell research, or research into the effects of certain kinds of drugs such as heroin (Frickel et al., 2010). Frickel et al. note that civil society groups may also play an important role in getting a social problem framed in a particular way. For instance, the success of Alcoholics Anonymous to frame the treatment of alcoholism as the lifelong abstinence from consuming alcohol effectively prevented researchers from conducting controlled drinking trials. 'The mere threat,' write Frickel et al. 'of interference from the grassroots was enough to keep many researchers from conducting certain studies. Several drug and alcohol researchers described great unwillingness to conduct studies on the health benefit of "harm reduction" programs, such as those that distribute free condoms in schools or clean needles in neighborhoods, because they might attract unwanted controversy from lay groups who oppose such public health interventions' (ibid.). This suggests the importance of examining the role a moral economy has in shaping what research a scientist will pursue and not pursue (Kohler, 1994). All of this points, again, to the 'uneven distribution of power and resources in science at the center' of sociology of science analyses (Frickel et al., 2010).

Thus, the progress in democratizing science by acknowledging and working with stakeholders beyond the academy and industry is not without issues. Joly and Kaufman's (2008) empirical research on knowledge mobilization about the issue of nanotechnology found engagement much more challenging than traditional knowledge translation because of the already-embedded alignment of powerful actors and a worldwide agenda with social-technical networks. Similarly, in their analysis of the public engagement initiative *GM Nation?*, which sought a public debate about transgenic (genetically modified) crops in the UK, Rowe et al. (2005) reported the additional challenge of completing a comprehensive evaluation of public engagement initiatives. These challenges included the difficulties in determining how we *should* (normative), and how we *do* (practical) evaluate public engagement strategies.

Industries and corporations, as we would expect, have themselves keyed into participation as a way of allaying public ambivalence about genetic databases, such as the UK Biobank (Tutton, 2007). Thus citizen science may be co-opted by those in positions of power, such as scientists, industry, corporations, and government. Dominique Pestre makes the important point that 'in our marked-based democracies, dialogic and participatory democracy is not central to the regulation of technoscience, techno-scientific knowledge and products. . . . Democracy is not a political regime free from conflict; discourses of participation have become central elements of a new form of governmentality' (2008: 101).

Gwen Ottinger's (2010) case study of a community's use of buckets to monitor air toxins found that the actual effectiveness—as opposed to expert rhetoric encouraging participatory citizen science—of lay involvement pivoted around issues of standards and standardized practices. Standards, Ottinger found, serve as both a 'boundary-bridging function that affords bucket monitoring data a crucial measure of legitimacy among experts. On the other hand, standards simultaneously serve a boundary-policing function, allowing experts to dismiss bucket data as irrelevant to the central project of air quality assessment' (ibid., 244). By adopting a scientific measure of air toxicity (the bucket method), in other words, the community group was able to claim legitimacy for their environmental concerns, but this legitimacy only went so far, because the scientific experts used standardization to preclude the community from full participation (because they were not scientists, the argument went, they could not produce as accurate or valid data as the scientists themselves). And bringing the discussion back to journalism, a number of studies examine journalists' roles in aiding corporate and special interest groups in manufacturing doubt about scientific research (see, for example, Stocking and Holstein, 2009).

What these sociological analyses highlight, and as Chapter 5 examined, is that there is a potential conflict of interest between expertise and democracy. Put another way, the social contract between science and society has shifted due to *both* the commercialization of science and greater opportunities for public input into scientific research programs (Nowonty, Scott, and Gibbons, 2001; Remington, 1988). Science and technology are clearly commercialized in contemporary society, with explicit and often implicit links with industries, corporations, and the military. Industry's entire enterprise is concerned with increasing profit. Thus, any discussion of science in society needs to take seriously questions about challenges to democratizing science. These challenges include intellectual property rights (who owns the copyright on mouse and human genes for instance), the World Wide Web and the digital divide, biotechnology and agriculture (GM foods, agribusiness), the politics of indigenous resources (bioprospecting, biocolonialism), gender and race in science and technology, weapons of mass destruction, environmental sustainability, and generally the possibilities and limitations of scientific citizenship in democratic states (Bell, 2006).

Science and Technology Close to Home

This book closes with an example of science, trust, values, and engagement that is, perhaps literally, closest to home. It concerns how we understand our deepest and most enduring roots through kinship. Kinship concerns how we understand ourselves as individuals, and how we understand our relationships with relations and non-relations. Kinship brings science, technology, and values into sharp relief (see Hird, 2009).

Because science and technology are social relations, science and technology and values co-produce each other (Bell, 2006; Hanks, 2010). Science will continue to both reflect old values, and produce new ones. Society's values will continue to inform and often direct science.

Kinship is most often defined within western society as either blood or non-blood relations. Blood relations are assumed to share biological substance including genes and blood, and to have resulted from sexual reproduction. The study of kinship has been one of the great mainstays of anthropology, and during the heyday of the anthropological tradition of extensive ethnographic study of non-western cultures, anthropologists discovered that some cultures used classificatory blood kinship terms that did not correspond to what were thought by Euro-North American anthropologists to be true genetic relationships (that is, biological). Trobriand Islanders and Aboriginals of Australia, for instance, deployed a complex system of relations to define kin, some based on what Euro-North American anthropologists recognized as blood relations, and some based on non-blood relations. In effect, Trobriand Island and Aboriginal systems of kinship challenged Euro-North American assumptions about the consanguinity of kinship. As Franklin observes, 'it was a perception that derived from the European scientific assumption that kinship categories should be read directly from "blood" ties as a matter of commonsense, and that to do otherwise could only be interpreted as ignorance of paternity, or general lack of intellectual development' (1997: 22). In his famous studies of the Trobriand Islanders, for instance, Bronislaw Malinowski argued, 'it seems hardly necessary to emphasize that for physiological consanguinity as such, pure and simple, there is no room in sociological science' (1913: 177fn.).

David Schneider analyzed the ways in which American culture has become increasingly dependent upon notions of biology. In his path-breaking work on kinship in *American Kinship: A Cultural Account* (1968), Schneider offered a sustained account of the complex relationship between biology and kinship.

Just 12 years later the hegemony of biology had become such that, in the second edition of the book, Schneider argues:

> In American cultural conception, kinship is defined as biogenetic. This definition says that kinship is whatever the biogenetic relationship is. If science discovers new facts about biogenetic relationship, then that is what kinship is and was all along. (1980: 23)

Because science is social relations, biology has no meaning outside its cultural context, and Schneider highlights the particular contradictions of Euro-North American understandings of kinship. He argues:

> The relationship between man [sic] and nature in American culture is an active one . . . Man's place is to dominate nature, to control it, to use nature's powers for his own ends . . . In American culture man's fate is seen as one which follows the injunction Master Nature! . . . But at home things are different. Where kinship and the family are concerned, American culture appears to turn things topsy-turvy. . . . What is out there in Nature, say the definitions of American culture, is what kinship is. . . . To be otherwise is unnatural, artificial, contrary to nature. (1968: 107)

Schneider's work has been the subject of critical analyses that have pointed out, for instance, that it relies upon a distinction between cultural facts and biological facts at the same time that it seeks to expose this distinction in other anthropological work (Franklin, 1997). Nevertheless, Schneider's focus on heterosexual sex as the central symbolic universe of American kinship has been taken up within contemporary lesbian, gay, and reproductive technology kinship studies. Franklin notes 'amid the many transformations that have reshaped the study of kinship over time, the question of the significance of biological facts has remained a persistent quagmire—as easy to fall into as it is difficult to leave behind' (2001: 302).

Reproductive technologies—a technology of the twentieth and twenty-first centuries—offer a number of challenges to the assumption that heterosexual coitus leads to pregnancy, which leads to offspring of direct kin relation to her/his parents. In the first instance, as Franklin observes, for the 'growing number of couples . . . [for whom] coitus *never* results in pregnancy, or for whom even conception and implantation do not result in pregnancy, the usefulness of the biological model is . . . in question' (1997: 64). For sub-fertile or infertile heterosexual individuals, coitus very rarely results in pregnancy. Moreover, in the majority of cases where reproductive technologies are used, conception and implantation of embryos also does not result in pregnancy. So right from the start, traditional understandings of kinship fall far short of the lived experiences of many heterosexual people, as well as lesbian and gay people. Add to this the growing use of sperm and egg donation, and the traditional understanding of kinship is further challenged.

Reproductive technologies introduce further challenges. A woman who uses egg donation might gestate and give birth to a child she has no genetic relationship with (or more specifically, no genetic relationship through the egg—genetic material does transfer through blood). Or a woman who uses the egg of her own mother might give birth to a child who is, genetically speaking, her sister. The list of variations goes on (see Thompson, 2001). In each of these cases, we might argue that kinship is *extended* beyond traditional criteria to include more than the person who gives birth to a child and her partner (i.e., to include egg donor, sperm donor, and so on).

Carlos Novas and Nikolas Rose observe that 'new reproductive technologies have split apart categories that were previously coterminous—birth mother, psychological mother, familial father, sperm donor, egg donor and so forth—thus transforming the relations of kinship that used to play such a fundamental role in the rhetoric and practices of identity formation' (2000: 490–1). As Charis Thompson observes, 'biological motherhood is becoming something that can be partial' (2001: 175). That is, reproductive technologies invite such emotive concern from the public because these technologies demonstrate that biogenetics underdetermines kinship, insofar as kinship is defined as both primordial and immutable. In this way, just as anthropologists found that so-called primitive cultures use classificatory systems, we could well argue that western cultures use these same classificatory systems, even whilst they depend upon strong notions of biology. That is, we assume that mother and child are blood related, that children do not share germ cells with their never born siblings. But these common assumptions may not be corroborated by biological evidence.

Recent scientific research investigating chimerism and mosaicism further confound traditional notions of embodiment (Hird, 2004a). Recent studies in biology refer to chimerism as the presence of two genetically distinct cell lines in an organism. This may occur through inheritance, transplantation, or transfusion. For instance, a boy was recently born in Britain who is, genetically speaking, two people because he was formed by the fertilization of two eggs and two sperm, which then fused into one embryo (Pearson, 2002). Cells traffic between fetus and mother in both directions during pregnancy, and those fetal cells continue to circulate for years in the mother after birth. This microchimerism has also been found in multiply transfused recipients of blood transfusions (Nelson, 2002).

Some interesting examples have turned up in the medical literature. The cell and tissue blood of one boy had none of his father's chromosomes, but did have a duplicated set of one half of his mother's chromosomes (Pearson, 2002). In another case, a mother was discovered not to be the genetic mother of her four children (whom she had gestated and given birth to, and had not used donor eggs). This woman has two populations of genetically different cells, one in her blood and the other in her gonads, and only the cells in her gonads were transferred to her children. Mosaicism is more common than chimerism and refers to patches of tissue that differ genetically. This would result in a person having two genetically distinct cell lines on a part or parts of their body.

Like reproductive technologies, chimerism and mosaicism challenge assumptions made about kinship. In some cases chimerism and mosaicism produce a similar extension of kinship criteria—for instance, to never-living siblings. But sometimes they produce the opposite—these biological variations *contract* kinship such that a woman who uses her own egg, uterus, and blood to produce a child might not be blood related or genetically related to this child. A man whose sperm is used to fertilize an egg that produces a child may not be blood related or genetically related to this child.

What is so interesting about chimerism and mosaicism is that whereas public understandings of reproductive technologies are deeply imbued with concerns about tampering with nature, chimerism and mosaicism are natural in the sense that they have undergone no human technological intervention (except in cases of transfusion or transplantation). Chimerism and mosaicism may be viewed as anomalies but they stand outside human technological intervention even as they fundamentally challenge traditional notions of kinship. As Franklin observes, 'ideas of the natural comprise one of the most important "cultural logics" that more recent theorists of kinship and gender have sought to analyze' (1997: 57). And in analyzing this cultural logic we find that nature and science are deployed in uncomfortably contradictory ways. What chimerism and mosaicism demonstrate is that science contradicts the cultural assumption that children are biologically related to their (non-adoptive) parents, at the same time that this cultural assumption is supposed to be grounded in biological explanation. It is for this reason that Franklin and McKinnon (2001) argue that privileging kinship rests on a tautology.

Moreover, science, as expert producer and interpreter of what is natural, is imbued with characteristics of rationality and impartiality within western traditions. Science may also reveal relationships where none are assumed (between living and never living siblings), and no relationship (between mother and child) where such a relationship is the foundation of kinship systems.

Processes of inclusion and exclusion are at the heart of cultural configurations of kinship. For instance, Kath Weston asks:

If kinship can ideologically entail shared substance, can transfers of bodily substance create—or threaten to create—kinship? Can they create—or threaten to create—other forms of social responsibility? What investment do people have in depicting the transfer of blood, organs, and sperm as sharing, giving or donation? What investment do they have in resisting such transfers (or the vehicles of transfer)? Alternatively, how do people work to construe transfers as 'signifying nothing' with respect to race, sexual contact, religious identity and so on? (2001: 153)

Just as reproductive technologies threaten established understandings of kinship (of inclusion and exclusion), Kath Weston argues science and technology offer both the promise *and* threat of new configurations of selfhood, responsibility, and kinship.

Prior to this research, chimerism was already courting centre stage within the context of research on xenotransplantation. At the launch of the joint report on xenotransplantation grafting by the British Union for the Abolition of Vivisection and Compassion in World Farming, one scientist warned, 'The human xenotransplantation patient will become a literal *chimera* . . . It sounds like scare-mongering, but let me assure you that the word *chimera* is being used by xenotransplantation scientists. . . .' (quoted in Brown, 1999a: 191; emphasis added).

Xenotransplantation involves the use of nonhuman animal cells or organs in human animals. We may think of the concept of kinship not only in terms of intra-species inclusions and exclusions (as in the case of human animals) but also between species. Thus Weston's questions about the boundaries of kinship do not just apply to the transfers of human organ and tissue between humans, but extend to these transfers between human and nonhuman animals.

A great deal of boundary work is done to continually distinguish between human and nonhuman animals (Hird, 2006). Xenotransplantation engenders public concern to the extent that it threatens to collapse notions of interspecies kinship boundaries. Haraway (1997) explores this boundary collapse in her work on biogenetic relationships such as OncoMouse™ that create a specific form of genetic relationship between humans and mice (humans and mice are, of course, already genetically related; so are humans and bananas, for that matter). These trans relationships 'simultaneously fit into well-established taxonomic and evolutionary discourses [for instance, technological progress] and also blast widely understood senses of the natural limit' (Haraway, 1997: 56). Franklin notes the way in which humans are today connected and related through biology *'undoes the very fixity that the biological tie used to represent'* (2001: 314).

Debates invoked by xenotransplantation are heavily dependent upon an implicit notion of the monster, in this case in terms of the authentic boundaries of the self (especially here in terms of humans versus nonhumans). The pollution created by this monstrous transgression of boundaries requires action: 'the delineation of a border, the naming of transgressors, the ritual of the purge, the subsequent restoration of a boundary' (Brown, 1999b: 342).

Thus public concerns about chimerism through xenotransplantation and reproductive technologies can be understood as contemporary distillations of kinship boundary work. Xenotransplantation and reproductive technologies effectively extend traditional understandings of kinship as flesh and blood. And both technologies do so through an explicit and primary use of notions of nature and science; the very same notions the western concept of kinship has relied upon to define (through exclusion) itself.

Summary

This chapter concerned the shifting social contract between science, technology, and society. Its main argument is that science operates within a contemporary context in which industrial elites increasingly influence if not determine entire scientific research programs from what gets studied (how social problems are defined), how things are studied, and to what ends scientific research is put. Science *also* operates, however, within a context of increased public engagement with science, also influencing research agendas.

The chapter began by considering the complicated relationship between good and bad science. The point here is that this is not a straightforward delineation, and we are faced with complex and uneasy questions about the relationship between values and research. How, for instance, should we as a society approach science (Nazi science conducted in the concentration camps is the most notorious example; see also Box 5.4) that caused untold suffering and also increases our knowledge and is put to the good end of helping people? There are many other considerations. For instance, scientific budgets are finite and Canadians must decide how to allocate financial resources. Canadians devote large sums of money to health research because Canadians strongly value health. Within the broad field of health research, which specific research programs deserve priority is complicated. For instance, given finite budgets, should money be directed to research that focuses on cancer prevention or research that treats cancer as a chronic disease (CBC, 2009)? How, as well, do we predict the ends of scientific research? Is weapons research ethically viable? How do we mediate the positive and negative impacts of science? For instance, interest in the opening up of the Northwest Passage (due to the melting of the polar ice cap because of global heating) has implications for both environmental sustainability as well as the economic boom forecast by neo-capitalists interested in the expansion of viable trade routes that may bring a higher standard of living for Inuit peoples (Vardy, 2009). The inherent uncertainty of scientific research profoundly undermines the ideal of modern science (Nowotny, Scott, and Gibbons, 2001).

The discussion of science and values brings to the fore the distinction between the ideal of science, and its actual practice as social relations. Merton's scientific norms point not so much to the ways in which ideal science may be compromised in particular contexts because this would suggest ideal science is actually possible to achieve. The point of this book is to argue science is indelibly social as humans, perceptions, machines, government funding, and a whole host of entities intra-act to produce scientific knowledge (see Chapter 3).

The degree to which publics trust science is also mediated by numerous factors including media coverage of scientific research, and most important, the social context within which individuals are situated. Sociological research suggests people process, circulate, and take up scientific knowledge in different ways depending on social factors such as class. This raises the issue of how science *wants* to be understood. The model used by science has shifted from attempting to increase the public understanding of science through knowledge translation techniques, to a focus on knowledge mobilization and citizen science. The former is concerned with the so-called faithful transmission of knowledge from experts to lay people. The latter recognizes a more distributed expertise, whereby different groups have different—and sometimes overlapping—expertise. Citizen science recognizes the rights of members of the public to have an active ongoing say in science. It is contrasted with the already embedded but largely invisible hand that industry elites have in determining science's research agenda. This influence is transparent enough, however, that science is no longer able to assert claims about autonomy or its inherent universalism or objectivity. These are, well and truly, contested claims. And interestingly, as expertise becomes more widely distributed, 'trust becomes an even more scarce and precious resource' (Nowotny, Scott, and Gibbons, 2001: 261).

The chapter closes with a consideration of science close to home. The discussion of scientific research on kinship brings to the fore complex issues with regard to identities as members of particular groups (families, races, genders), and the profound part science and technology play in shifting our conceptualization of where we come from, who we are, and to which groups we belong. Scientific research on kinship ties science to social relations in profound material and symbolic social relations. It also brings to the fore the reality that science will continue to proliferate, rather than eradicate, uncertainties (Nowotny, Scott, and Gibbons, 2001). The challenge for science and society is to develop ways of coping with these ongoing uncertainties in ways that emphasize social justice and the democratization of knowledge.

Key Terms

Royal Society   Possibly the oldest western learned society for the study of science, the Royal Society is located in London, United Kingdom.

Norm   Social norms refer to expected patterns of behaviour. Sociologists analyze the ways in which norms are created and maintained within different cultures.

Scientific literacy   This term refers to the knowledge and understanding about material phenomena and scientific practices.

Deficit model   This term refers to the supposition that individuals do not sufficiently understand science and nature. Some governments have created policies and programs designed to increase their constituents' knowledge about science and nature.

Visible Human Project   This is a project to detail the anatomy of the human body through detailed cross-sectional photographs of the human body.

Democracy   This term refers to a form of government whereby governing decisions are made by the people either through referenda or elected representatives.

Kinship   This term refers to the relationship between two or more entities. The term is typically used to define the relationship between individual people through a combination of biological (such as genes) and cultural (such as being raised in the same family) factors.

Chimerism   This term describes the existence of two distinct populations of cells (different genotypes) within a single organism.

Mosaicism   This term describes the existence of two distinct populations of cells arising from the same zygote within a single organism.

Xenotransplantation   This term refers to the transplantation of living cells from one species to a different species, such as from pigs to humans.

Critical Thinking Questions

1. Scientists, the Canadian government, and Canadian citizens need to ask questions about values in any given scientific research program, such as What are the benefits? What are the costs? Who benefits? and Who incurs risk? What other questions should be asked?

2. Should members of the public have a say in what specific research projects scientists undertake?

3. Do you trust science?

4. If the goal of scientific research is to create a better society, do the ends justify the means? What about patenting human genes in which individual persons lose the legal right over their own genes?

Suggested Readings

M. Burawoy, '2004 ASA Presidential Address: For Public Sociology', *The British Journal of Sociology* 56, 2 (2005): 259–94. This article details a well-known sociologist's ideas about how sociology might better engage with society in the future.

J. Chilvers, 'Deliberating Competence: Theoretical and Practitioner Perspectives on Effective Participatory Appraisal Practice', *Science, Technology, and Human Values*

33, 2 (2008): 155–85. This article provides a useful and timely analysis of the opportunities and challenges of participatory science approaches.

B. Cooke and U. Kothari, *Participation: The New Tyranny?* London: Zed Books, 2001. This book provides a thought-provoking critique of participatory models of citizen science, with a particular focus on labour relations and development.

B. Latour, 'From Realpolitik to Dingpolitik or How to Make Things Public', in B. Latour and P. Weibel, eds, *Making Things Public: Atmospheres of Democracy*. Cambridge, MA: MIT Press, 2005, pp. 14–41. This chapter is another thought-provoking critique from well-known sociologist Bruno Latour and Peter Weibel, the Director of ZKM, Center for Art and Media Technology. The article appears in a book compiling over one hundred well-known science and technology scholars interested in the intersections between science and politics. The chapter asks what it means to make *things* public.

J. Petts and C. Brooks, 'Expert Conceptualizations of the Role of Lay Knowledge in Environmental Decision-making: Challenges for Deliberative Democracy', *Environment and Planning A* 38 (2006): 1045–59. This article focuses on how willing experts are to accommodate and/or change scientific practices based on lay input concerning environmental issues.

P. Reason and H. Bradbury, eds, *The SAGE Handbook of Action Research: Participative Inquiry and Practice*, 2nd edn. Los Angeles: SAGE Publications, 2008. Action research prioritizes the concerns, priorities, and ways of knowing that non-experts bring to a given research project. This book provides a comprehensive overview of the literature, including the benefits and challenges, of this type of research.

G. Rowe and L. Frewer, 'A Typology of Public Engagement Mechanisms', *Science, Technology, and Human Values* 30 (2005): 251–90. This article provides an overview of the main ways in which publics engage with societal issues involving science and technology.

M.A. Saari, C. Gibson, and A. Osler, 'Endangered Species: Science Writers in the Canadian Daily Press', *Public Understanding of Science* 7 (1998): 61–81. This article argues that the profession of science writing—non-scientists who report on scientific research for lay audiences—is under threat due to changing reporting priorities.

P. Stassart, *Running an Interdisciplinary Competency Group*. Centre for Rural Economy Discussion Paper, 2008. Available at www.ncl.ac.uk/cre/publish/discussionpapers/pdfs/dp19.pdf. This report provides a review of a hands-on attempt to organize and implement a science and technology competency group. The paper's details about the challenges faced when researchers try to get people from different disciplines to talk with each other and work on a particular issue of concern together, is particularly useful.

S. Whatmore, 'Mapping Knowledge Controversies: Science, Democracy and the Redistribution of Expertise', *Progress in Human Geography* 33 (2009): 587–98. In this article, Sarah Whatmore describes the competency group that she and her team at Oxford University put together to help address the recurrent issue of flooding in parts of the United Kingdom. Whatmore's analysis of the opportunities and challenges of the competency group approach is a very useful contribution to thinking about the potentials of citizen science.

Websites and Films

Oxford University Centre for the Environment
www.ouce.ox.ac.uk
> This website provides information about an innovative research group that brings together a diverse range of social and natural scientists, as well as members of the public, industry, and government, to addressing pressing environmental issues.

The Chernobyl Disaster
www.world-nuclear.org/info/chernobyl/inf07.html
> This website provides thought-provoking and at times disturbing evidence of the ramifications of the nuclear power plant accident in Ukraine. The website is sponsored by the World Nuclear Organization so it is pro-nuclear power. Students should view this website from a critical sociological perspective as an example of how scientists and industry present issues of concern, such as environmental safety, to the public.

Medical Experiments of the Holocaust and Nazi Medicine
http://remember.org/educate/medexp.html
> This website provides an overview of the 'experiments' performed on concentration camp prisoners during the Second World War. The descriptions are disturbing and they beg salient questions about the role of ethics in science and society.

Body Worlds
www.bodyworlds.com/en.html
> Body Worlds is a show that tours several countries, presenting plastinated human and other animal bodies for people to look at. The aim of the show, according to its creator Gunther von Hagens, is to invite reflection upon what it means to be human in a modern scientific age.

Canadian Research Ethics
www.researchethics.ca
> This website provides a series of publications and links concerned with issues related to research ethics.

Uranium
> This film, produced by the National Film Board of Canada, focuses on the health, safety, and environmental hazards of uranium mining in Canada. The film provokes important questions about how Canadians can and should be involved in making provincial and national policies on issues like uranium mining, nuclear power, and so on.

Crapshoot
> This National Film Board of Canada film focuses on what happens to the garbage produced by Canadians. It especially looks at the ways in which garbage becomes part of landfills and ends up in our water systems via sewage pipes. The film invites critical sociological questions about how our consumerist society creates waste, and how we deal with it, as well as how Canadian citizens might better participate in citizen science in order to become better informed, and have a greater say in how our garbage is managed.

References

Aaboe, A. 1974. 'Scientific Astronomy in Antiquity', *Philosophical Transactions of the Royal Society* 276 (1257): 21–42.

———. 1991. 'The Culture of Babylonia: Babylonian Mathematics, Astrology, and Astronomy', in J. Boardman, I. Edwards, N. Hammond, E. Sollberger, and C. Walker, eds, *The Assyrian and Babylonian Empires and other States of the Near East, from the Eighth to the Sixth Centuries B.C.* Cambridge: Cambridge University Press.

Acton, C., and M. Hird. 2004. 'Toward a Sociology of Stammering', *Sociology* 38 (3): 495–513.

Adams, C. 1995. *Neither Man Nor Beast: Feminism and the Defense of Animals.* New York: Continuum.

Agamben, G. 2003. *The Open: Man and Animal.* Chicago: University of Chicago Press.

Agyeman, J., P. Cole, and R. Haluza-Delay, eds. 2009. *Speaking for Ourselves: Environmental Justice in Canada.* Vancouver: University of British Columbia Press.

Ainley, M. 1990. *Despite the Odds.* Montreal: Vehicule Press.

Alexander, A.R. 2002. *Geometrical Landscapes: The Voyages of Discovery and the Transformation of Mathematical Practice.* Stanford, CA: Stanford University Press.

Allchin, D. 1998. 'Values in Science and in Science Education', in B.J. Fraser and K.G. Tobin, eds, *International Handbook of Science Education.* Dordrecht: Kluwer Academic Publishers, pp. 1083–92.

Altmann, J. 1980. *Baboon Mothers and Infants.* Cambridge, MA: Harvard University Press.

American Anthropological Association. 1998. 'Statement on "Race"'; available at www.aaanet.org/stmts/racepp.htm.

American Behavioral Scientist, 37(6).

Anderson, W. 2006. *The Cultivation of Whiteness: Science, Health, and Racial Destiny in Australia.* Durham, NC: Duke University Press.

Armstrong, R., E. Waters, B. Crockett, and H. Keleher. 2007. 'The Nature of Evidence Resources and Knowledge Translation for Health Promotion Practitioners', *Health Promotion International* 22 (3): 254–60.

Aronowitz, S. 1988. *Science as Power: Discourse and Ideology in Modern Society.* Minneapolis, MN: University of Minnesota Press.

Auletta, G. 2006. 'Critical Examination of the Conceptual Foundations of Classical Mechanics in the Light of Quantum Physics'; available at http://cdsweb.cern.ch/record/490325/files/0103047.pdf.

BBC News. 2009. 'Colleague Defends "ClimateGate" Professor', *BBC News Online* (4 December). Retrieved from http://news.bbc.co.uk/2/hi/8396035.stm

BBC News. 2010. 'Climategate e-mails inquiry under way', *BBC News Online* (11 February). Retrieved from http://news.bbc.co.uk/2/hi/8510498.stm

Bacon, F. [1620] 2000. *The New Organon*, L. Jardine and M. Silverthorne, eds. Cambridge: Cambridge University Press.

Bagemihl, B. 1999. *Biological Exuberance, Animal Homosexuality and Natural Diversity.* New York: St. Martin's Press.

Barad, K. 2001. 'Scientific Literacy → Agential Literacy = (Learning + Doing) Science Responsibly', in M. Mayberry, B. Subramaniam, and L. Weasel, eds, *Feminist Science Studies: A New Generation.* New York: Routledge Press, pp. 226–47.

———. 2007. *Meeting the Universe Halfway: Quantum Physics and the Entanglement of Matter and Meaning.* Durham, NC: Duke University Press.

Barnes, B. 2005. 'The Credibility of Scientific Expertise in a Culture of Suspicion', *Interdisciplinary Science Reviews* 30 (1): 11–18.

Barry, A., G. Born, and G. Weszkalnys. 2008. 'Logics of Interdisciplinarity', *Economy and Society* 37 (1): 20–49.

Bauchspies, W., J. Croissant, and S. Restivo. 2006. *Science, Technology, and Society: A Sociological Approach.* London: Blackwell Publishing.

Bauer, M. 2007. 'The Public Career of the "Gene"—Trends in Public Sentiments from 1946 to 2002', *New Genetics and Society* 26 (1): 29–45.

Bauer, M., and I. Schoon. 1993. 'Mapping Variety in Public Understanding of Science', *Public Understanding of Science* 2: 141–55.

Becker, G. 1992. 'The Merton Thesis: Oetinger and German Pietism, a Significant Negative Case', *Sociological Forum* 7 (4): 641–60.

Behe, M.J. 1996. *Darwin's Black Box: The Biochemical Challenge to Evolution.* New York: Free Press.

Bell, D. 2006. *Science, Technology and Culture.* Milton Keynes: Open University Press.

Berger, C. 1996. *Honour and the Search for*

Influence: A History of the Royal Society of Canada. Toronto: University of Toronto Press.

Bergson, H. 1911. *Matter and Memory*, Nancy Margaret Paul and W. Scott Palmer, trans. London: George Allen and Unwin.

Birke, L. 1994. *Feminism, Animals and Science: The Naming of the Shrew*. Toronto: McGraw Hill.

Birke, L., M. Bryld, and N. Lykke. 2004. 'Animal Performances: An Exploration of Intersections Between Feminist Science Studies and Studies of Human/Animal Relationships', *Feminist Theory* 5 (2): 167–83.

Bloor, D. 1999. 'Anti-Latour', *Studies in History and Philosophy of Science* 30 (1): 113–29.

Blume, S. 1974. *Toward a Political Sociology of Science*. New York: Free Press.

Bohr, N. 1961. *Atomic Physics and Human Knowledge*. New York: Science Editions, Inc.

Bothwell, R. 1988. *Nucleus: The History of Atomic Energy of Canada Limited*. Toronto: University of Toronto Press.

Braidotti, R. 2006. *Transpositions*. Cambridge: Polity Press.

Briggs, L. 2002. *Reproducing Empire: Race, Sex, Science, and U.S. Imperialism in Puerto Rico*. Berkeley, CA: University of California Press.

Broadhurst Dixon, J., and E. Cassidy. 1998. *Virtual Futures*. London: Routledge.

Brown, N. 1999a. 'Debates in Xenotransplantation: On the Consequences of Contradiction', *New Genetics and Society* 18 (2/3): 181–96.

———. 1999b. 'Xenotransplantation: Normalizing Disgust', *Science as Culture* 8 (3): 327–55.

Brown, N., and M. Michael. 2001. 'Transgenics, Uncertainty and Public Credibility', *Transgenic Research* (10): 279–83.

Bryson, B. 2005. *A Short History of Nearly Everything: Special Illustrated Edition*. New York: Broadway Press.

Bucchi, M. 2004. 'Can Genetics Help Us Rethink Communication? Public Communication of Science as a "Double Helix"', *New Genetics and Society* 23 (3): 269–83.

Cajete, G. 2000. *Native Science: Natural Laws of Interdependence*. Santa Fe, New Mexico: Clear Light Publishers.

Callon, M., and B. Latour. 1992. 'Don't Throw the Baby out with the Bath School!', in A. Pickering, ed., *Science as Practice and Culture*. Chicago: Chicago University Press, pp. 343–68.

Callon, M., and J. Law. 1997. 'After the

Individual in Society: Lessons on Collectivity From Science, Technology and Society', *Canadian Journal of Sociology* 22 (2): 1–11.

Canadian Institutes of Health Research. 2004. 'Knowledge Translation Strategy 2004–2009'; available at www.cihr-irsc.gc.ca/e/26574.html.

Cartwright, N. 1983. *How the Laws of Physics Lie*. Oxford: Oxford University Press.

Cavell, S. 1988. *In Quest of the Ordinary: Lines of Skepticism and Romanticism*. Chicago: University of Chicago Press.

CBC. 2009. 'Cancer as a Chronic Disease', *Quirks and Quarks*, 30 October.

Clark, N. 2008. 'Aboriginal Cosmopolitanism', *International Journal of Urban and Regional Research*, 1–8.

———. 2011. *Inhuman Nature: Sociable Life on a Dynamic Planet*. London and New York: Sage Publications.

Cohen, I.B. 1980. *The Newtonian Revolution: With Illustrations of the Transformation of Scientific Ideas*. Cambridge: Cambridge University Press.

Cohen, M.M. 2010. 'Overview of German, Nazi, and Holocaust Medicine', *American Journal of Medical Genetics Part A* 152A (3): 687–707.

Collins, H. 1974. 'The TEA Set: Tacit Knowledge and Scientific Networks', *Science Studies* 4 (2): 165–86.

———. 1991. *Changing Order: Replication and Induction in Scientific Practice*. Chicago: The University of Chicago Press.

Collins, H., and R. Evans. 2002. 'The Third Wave of Science Studies: Studies of Expertise and Experience', *Social Studies of Science* 32: 235–96.

Collins, H., and T. Pinch. 1993. *The Golem: What Everyone Should Know About Science*. Cambridge: Cambridge University Press.

Collins, H.M., and S. Yearley. 1992. 'Epistemological Chicken', in A. Pickering, ed., *Science as Practice and Culture*. Chicago, Chicago University Press, pp. 301–26.

Comte, A. [1853] 2009. *The Positive Philosophy of Auguste Comte*, 2 volumes. H. Martineau, trans. Cambridge: Cambridge University Press.

Condit, C., R. Parrot, T. Harris, J. Lynch, and T. Dubriwny. 2004. 'The Role of "Genetics" in Popular Understandings of Race in the United States', *Public Understanding of Science* 13: 249–72.

Cooper, R. 2003. 'Race, Genes, and Health—New Wine in Old Bottles?', *International Journal of Epidemiology* 32: 23–5.

Copeland, H. 1956. *The Classification of Lower Organisms*. Palo Alto: Pacific Books.

Cosmides, L., and J. Tooby. 1992. 'Cognitive Adaptations for Social Exchange', in J. H. Barkow, L. Cosmides, and J. Tooby, eds, *The Adapted Mind: Evolutionary Psychology and the Generation of Culture*. New York: Oxford University Press.

Crick, F. 1958. 'Central Dogma of Molecular Biology', *Nature* 227: 561–3.

Cummings, S. 1984. 'The Political Economy of Social Science Funding', *Sociological Inquiry* 54 (2): 154–70.

Curtis, T., W. Sloan, and J. Scannell. 2002. 'Estimating Prokaryote Diversity and its Limits', *Proceedings of the National Academy of Sciences USA* 99: 10494–9.

Daly, M., and M. Wilson. 1988. *Homicide*. New York: Aldine De Gruyter Press.

Darwin, C. [1839] 1972. *The Voyage of the Beagle*. New York: Bantam Books.

———. [1859] 1998. *The Origin of Species*. Ware, Hertfordshire: Wordsworth Editions Limited.

———. [1871] 2004. *The Descent of Man, and Selection in Relation to Sex*. London: Penguin Classics.

Daston, L. 2007. Interview on *How To Think About Science*. CBC Series.

Daston, L., and P. Galison. 2007. *Objectivity*. New York: Zone Books.

Daston, L., and K. Park. 1998. *Wonders and the Order of Nature*. New York: Zone Books.

Davis, D., M. Evans, A. Jada, L. Perrier, D. Rath, D. Ryan, G. Sibbald, S. Straus, S. Rappolt, M. Wowk, M. Zwarenstein. 2003. 'The Case for Knowledge Translation: Shortening the Journey from Evidence to Effect', *British Medical Journal*, 327, 5 July: 33-35.

Davis, L. 1995. *Enforcing Normalcy: Disability, Deafness and the Body*. New York: Verso.

Davis, L., ed. 2006. *The Disability Studies Reader*, 2nd edn. London: Routledge.

Davis, M. 2001. *Late Victorian Holocausts: El Nino Famines and the Making of the Third World*. London and New York: Verso.

Dawkins, R. 1981. 'In Defense of Selfish Genes', *Philosophy* 56 (218): 556–73.

Degeling, C. 2009. 'Negotiating Value: Comparing Human and Animal Fracture Care in Industrial Societies', *Science, Technology, and Human Values* 34 (1): 77–101.

Dewalt, B. 1995. *Technology and Canadian Printing: A History from Lead Type to Lasers, Transformation Series 3*. Ottawa: National Museum of Science and Technology.

Di Chiro, G. 1998. 'Environmental Justice from the Grassroots', in D.J. Faber, ed., *The Struggle for Ecological Democracy*. New York: Guilford, pp. 104–36.

Dixon, H., M. Scully, M. Wakefield, and M. Murphy. 2009. 'The Prostate Cancer Screening Debate: Public Reaction to Medical Controversy in the Media', *Public Understanding of Science* 18: 115–28.

Donchin, A. 1989. 'The Growing Feminist Debate over the New Reproductive Technologies', *Hypatia* 4 (3): 136–49.

Donovan, J., and C. Adams, eds. 2000. *Beyond Animal Rights: A Feminist Caring Ethic for the Treatment of Animals*. New York: Continuum.

Duhem, P. 1954. *The Aim and Structure of Physical Theory*. Princeton, NJ: Princeton University Press.

Dummett, M. 1978. 'Realism', in *Truth and Other Enigmas*. Cambridge, MA: Harvard University Press, pp. 145–65.

Dunwoody, S. 1992. 'The Challenge for Scholars of Popularized Science Communication: Explaining Ourselves', *Public Understanding of Science* 1: 11–14.

Dupré, J. 1996. *The Disorder of Things: Metaphysical Foundations of the Disunity of Science*. Cambridge, MA: Harvard University Press.

Durant, J., G. Evans, and G. Thomas. 1989. 'The Public Understanding of Science', *Nature* 340: 11–14.

Durkeim, E. [1895] 1982. *Rules of Sociological Method*, S. Lukes, ed. New York: Free Press.

Ede, A., and L. Cormack. 2004. *A History of Science in Society: From Philosophy to Utility*. Peterborough, Ontario: Broadview Press.

Edmonds, A. 1973. *Voyages to the Edge of the World*. Toronto: McClelland and Stewart.

Edwards, P. 1976. 'A Classification of Plants into Higher Taxa Based on Cytological and Biochemical Criteria', *Taxon* 25 (5/6): 529–42.

Eggleston, W. 1978. *National Research in Canada: The NRC 1916–1966*. Toronto: Clarke Irwin.

Einsiedel, E., E. Jelsøe, and T. Breck. 2001. 'Publics at the Technology Table: The Consensus Conference in Denmark, Canada, and Australia', *Public Understanding of Science* 10: 83–98.

Einstein, B., B. Podolsky, and N. Rosen. 1935. 'Can Quantum-Mechanical Description of Physical Reality be Considered Complete?', *Physical Review* 47: 777–80, by the American Physical Society. Excerpt reprinted with permission.

Engelhardt, H.T., Jr., and A. Caplan. 1987, eds. *Scientific Controversies: Case Studies in the Resolution and Closure of Disputes in Science and Technology*. Cambridge:

Cambridge University Press.

Epstein, S. 1996. *Impure Science: AIDS, Activism, and the Politics of Knowledge*. Berkeley, CA: University of California Press.

———. 2007. *Inclusion: The Politics of Difference in Medical Research*. Chicago and London: The University of Chicago Press.

Farkas, N. 1999. 'Dutch Science Shops: Matching Community Needs with University R&D', *Science Studies* 12: 33–47.

Fausto-Sterling, A. 1997. 'Feminism and Behavioral Evolution: A Taxonomy', in P. Gowaty, ed., *Feminism and Evolutionary Biology*. New York: Chapman and Hall, pp. 42–60.

Featherstone, M., and R. Burrows. 1995. *Cyberspace, Cyberbodies, Cyberpunk*. London: Sage.

Fedigan, L. 1984. 'Sex Ratios and Sex Differences in Primatology', *American Journal of Primatology* 7: 305–8.

Ferreira, C. 2004. 'Risk, Transparency and Cover Up: Media Narratives and Cultural Resonance', *Journal of Risk Research* 7 (2): 199–211.

Feyerabend, P. [1975] 1993. *Against Method*. London: Verso.

Field, H., and P. Powell. 2001. 'Public Understanding of Science versus Public Understanding of Research', *Public Understanding of Science* 10: 421–6.

Flanagan, T. 2008. *First Nations? Second Thoughts*, 2nd edn. Montreal and Kingston: McGill-Queen's University Press.

Flannery, T. 1994. *The Future Eaters: An Ecological History of the Australian Lands and People*. New York: Grove Press.

Fleck, L. [1935] 1979. *Genesis and Development of a Scientific Fact*. Chicago: University of Chicago Press.

Foley, H., A. Scaroni, and C. Yekel. 2010. *RA-10 Inquiry Report: Concerning the Allegations of Research Misconduct Against Dr. Michael E. Mann, Department of Meteorology, College of Earth and Mineral Sciences, The Pennsylvania State University*. Available at www.research.psu.edu/orp/Findings_Mann_Inquiry.pdf.

Foucault, M. [1966] 2002. *The Order of Things*. London: Routledge Classics.

Franklin, A. 2008. 'A Choreography of Fire: A Posthumanist Account of Australians and Eucalypts', in A. Pickering and K. Guzik, eds, *The Mangle in Practice*. Durham, NC: Duke University.

Franklin, S. 1995. 'Romancing the Helix: Nature and Scientific Discovery', in L. Pearce and J. Stacey, eds, *Romance Revisited*. London: Lawrence and Wishard, pp.

63–77.

———. 1997. *Embodied Progress: A Cultural Account of Assisted Conception*. London and New York: Routledge.

———. 2000. 'Life Itself: Global Nature and the Genetic Imaginary' in Sarah Franklin , Celia Lury, and Jackie Stacey, *Global Nature, Global Culture*. London: Sage, pp.188–227.

———. 2001. 'Biologization Revisited: Kinship Theory in the Context of the New Biologies', in S. Franklin, and S. McKinnon, eds, *Relative Values: Reconfiguring Kinship Studies*. Durham, NC: Duke University Press, pp. 302–25.

———. 2007. *Dolly Mixtures: The Remaking of Genealogy*. Durham, NC: Duke University Press.

Franklin, S., and S. McKinnon, eds. 2001. *Relative Values: Reconfiguring Kinship Studies*. Durham, NC: Duke University Press.

Franklin, U. 1992. *The Real World of Technology*. Concord, ON: House of Anansi Press Ltd.

French, M., and M.J. Hird. 2008. *Concepts for Exploring Normative Structure in Scientific Communities: A Report on Potential Tools for Assessing the Problem Gambling Research Community*. Report prepared for The Structure of the Problem Gambling Research Community Research Project, OPGRC.

Frickel, S. et al. 2010. 'Undone Science: Charting Social Movement and Civil Society Challenges to Research Agenda Setting', *Science, Technology and Human Values* 35 (4); 444–473.

Fuglsang, L. 2005. 'IT and Senior Citizens: Using the Internet for Empowering Active Citizenship', *Science, Technology, and Human Values* 30 (4): 468–95.

Fuller, S. 1988. *Social Epistemology*. Bloomington, IN: Indiana University Press.

Gaard, G., ed. 1992. *Ecofeminism: Women, Animals, Nature*. Philadelphia: Temple University Press.

Genuis, S., and S. Genuis. 2006. 'Exploring the Continuum: Medical Information to Effective Clinical Practice. Paper 1: The Translation of Knowledge into Clinical Practice', *Journal of Evaluation in Clinical Practice* 12 (1): 49–62.

Ghiselin, M. 1997. *Metaphysics and the Origin of Species*. Albany, NY: State University of New York Press.

Gilbert, S. 2002. 'The Genome in Its Ecological Context: Philosophical Perspectives on Interspecies Epigenesis', *Ann. N.Y. Acad. Sci.* 981: 202–18.

Goggin, G., and C. Newell. 2004. 'Disabled E-Nation: Telecommunications, Disability, and National Policy', *Prometheus* 22 (4): 411–22.

Goldman, S.L. 2006. *The Science Wars: What Scientists Know and How They Know It.* Chantilly, VI: The Teaching Company, Para. 32; available at http://teachingcompany.12.forumer.com/viewtopic. php?t=1689.

Görke, A., and G. Ruhrmann. 2003. 'Public Communication between Facts and Fictions: On the Construction of Genetic Risk', *Public Understanding of Science* 12: 229–41.

Gould, S.J. 1981. *Evolution as Fact and Theory, The Unofficial Stephen Jay Gould Archive.* Available at www.stephenjaygould. org/library/gould_fact-and-theory.html.

Gould, S. 1996. *The Mismeasure of Man.* New York: W.W. Norton and Company.

———. 2000. *Wonderful Life. The Burgess Shale and the Nature of History.* London: Vintage.

Graham, I., J. Logan, M. Harrison, S. Straus, J. Tetroe, W. Caswell, and N. Robinson. 2006. 'Lost in Knowledge Translation: Time for a Map?', *The Journal of Continuing Education in the Health Professions* 26: 13–24.

Gray, C. 1995. *The Cyborg Handbook.* New York: Routledge.

Greene, S. 2004. 'Indigenous People Incorporated?: Culture as Politics, Culture as Property in Pharmaceutical Bioprospecting', *Current Anthropology* 45 (2): 211–237.

Griffiths, T. 2000. 'Traveling in Deep Time: La Longue Durée in Australian history', *Australian Humanities Review* 18 (June–August). Available at www.lib.latrobe. edu.au/AHR/archive/Issue-June-2000/griffiths4.html.

Gross, M. 2007. 'The Unknown in Process: Dynamic Connections of Ignorance, Non-Knowledge, and Related Concepts', *Current Sociology* 55: 742–59.

———. 2009. *Ignorance and Surprise: Science, Knowledge Production, and the Making of Robust Ecological Design.* Cambridge, MA: MIT Press.

Haber, L. 1986. *The Poisonous Cloud: Chemical Warfare in the First World War.* Oxford: Oxford University Press.

Hacking, I. 1983. *Representing and Intervening: Introductory Topics in the Philosophy of Natural Science.* Cambridge: Cambridge University Press.

———. 1999. *The Social Construction of What?* Cambridge, MA: Harvard University Press.

———. 2005. 'Why Race Still Matters', *Daedalus* 134 (1): 102–16.

Halder, G., P. Callaerts, and W. Gehring. 24 March 1995. 'Induction of Ectopic Eyes by Targeted Expression of the Eyeless Gene in Drosophila', *Science* 267: 1788–92.

Hamilton, W.D. 1964. 'The Genetical Evolution of Social Behavior', *The Journal of Theoretical Biology* 7: 1–16.

Hankinson-Nelson, L., and J. Nelson. 1996. *Feminism, Science, and the Philosophy of Science.* Dordrecht: Kluwer.

Hanks, C. 2010. *Technology and Values: Essential Readings.* Oxford: Wiley-Blackwell.

Haraway, D. 1989. *Primate Visions: Gender, Race, and Nature in the World of Modern Science.* New York: Routledge.

———. 1991. 'A Cyborg Manifesto: Science, Technology, and Socialist-Feminism in the Late Twentieth Century', in *Simians, Cyborgs and Women.* New York: Routledge, pp. 149–82.

———. 1992. 'The Promises of Monsters: A Regenerative Politics for Inappropriate/d Others', in L. Grossberg, C. Nelson, and P.A. Treichler, eds, *Cultural Studies.* New York: Routledge, pp. 295–337.

———. 2003. *The Companion Species Manifesto: Dogs, People, and Significant Otherness.* Chicago: Prickly Paradigm Press.

———. 1997. *Modest_Witness@Second_Millenium.FemaleMan_Meets_OncoMouse: Feminism and Technoscience.* New York and London: Routledge.

———. 1997. *Modest_Witness@Second_Millenium.FemaleMan_Meets_OncoMouse: Feminism and Technoscience.* New York and London: Routledge.

———. 2008. *When Species Meet.* Minneapolis, MN, and London: University of Minnesota Press.

Harding, S. 1998. *Is Science Multi-cultural? Postcolonialisms, Feminisms, and Epistemologies.* Bloomington, IN: Indiana University Press.

Harman, G. 2009. *Prince of Networks: Bruno Latour and Metaphysics.* Prahran, Vic.: re.press.

Harris, R. 1976. *A History of Higher Education in Canada: 1663–1960.* Toronto: University of Toronto Press.

Hartigan, J., Jr. 2008. 'Is Race Still Socially Constructed? The Recent Controversy Over Race and Medical Genetics', *Science as Culture* 17 (2): 163–93.

Heijmans B.T., et al. 4 November 2008. 'Persistent Epigenetic Differences Associated with Prenatal Exposure to Famine in Humans', *Proceedings of the National Academy of Sciences USA* 105 (44): 17046–9.

Heilbron, J., ed. 2003. *The Oxford Companion to the History of Modern Science*. Oxford: Oxford University Press.

Henrion, C. 1997. *Women in Mathematics: The Addition of Difference*. Bloomington: Indiana University Press.

Hess, D. 2009. 'The Potentials and Limitations of Civil Society Research: Getting Undone Science Done', *Sociological Inquiry* 79 (3): 306–27.

Hess, P., H.M. Preloran, and C. Browner. 2009. 'Diagnostic Genetic Testing for a Fatal Illness: the Experience of Patients with Movement Disorders', *New Genetics and Society* 28 (1): 3–18.

Heylighen, F. 1990. 'Classical and Non-Classical Representations in Physics', *Cybernetics and Systems* 21: 423–44.

Hickman, L. 20 November 2009. 'Climate Sceptics Claim Leaked Emails Are Evidence of Collusion Among Scientists', *The Guardian*, p. 19.

Hird, M.J. 2003. 'From The Culture of Matter to the Matter of Culture: Feminist Explorations of Nature and Science', *Sociological Research Online* 8 (1); available at www.socresonline.org.uk/8/1/hird.html.

———. 2004a. 'Chimerism, Mosaicism and the Cultural Construction of Kinship', *Sexualities* 7 (2): 225–40.

———. 2004b. *Sex, Gender and Science*. Houndmills, Basingstoke: Palgrave Press.

———. 2006. 'Animal Trans', *Australian Feminist Studies* 21 (49): 35–48.

———. 2009. *The Origins of Sociable Life: Evolution after Science Studies*. New York: Palgrave Macmillan.

Horst, M. 2007. 'Public Expectations of Gene Therapy: Scientific Futures and their Performative Effects on Scientific Citizenship', *Science, Technology, and Human Values* 32 (2): 150–71.

Houghtaling, M. Forthcoming. 'Toward an Existential Theory of Sexuality', PhD Dissertation, Queen's University.

Hrdy, S.B. 1974. 'Male–male Competition and Infanticide among the Langurs (*Presbytis entellus*) of Abus, Rajastan', *Folia Primatologica* 22: 19–58.

———. 1981. *The Woman That Never Evolved*. Cambridge, MA: Harvard University Press.

———. 1986. 'Empathy, Polyandry, and the Myth of the Coy Female', in R. Bleier, ed., *Feminist Approaches to Science*. New York: Pergamon Press, pp. 119–46.

———. 1997. 'Raising Darwin's Consciousness: Female Sexuality and the Prehominid Origins of Patriarchy', *Human Nature*

8: 1–49.

Hubbard, R. 1989. 'Science, Facts, and Feminism', in N. Tuana, ed., *Feminism and Science*. Bloomington, IN: Indiana University Press, pp. 119–31.

Hume, D. [1739] 2000. *A Treatise of Human Nature*. Oxford: Oxford University Press.

Huxley, J., and J. Baker. 1974. *Evolution: The Modern Synthesis*. New York: Allen and Unwin.

Hyde, J.S. 2005. 'The Gender Similarities Hypothesis', *American Psychologist* 60 (6): 581–92.

Irwin, A. 1995. *Citizen Science: A Study of People, Expertise and Sustainable Development*. London: Routledge.

———. 2001. *Sociology and the Environment: A Critical Introduction to Society, Nature and Knowledge*. Oxford: Polity Press.

———. 2006. 'The Politics of Talk: Coming to Terms with the "New" Scientific Governance', *Social Studies of Science* 36 (2): 299–320.

Irwin, A., and B. Wynne. 1996. *Misunderstanding Science? The Public Reconstruction of Science and Technology*. Cambridge: Cambridge University Press.

Isaacson, W. 2008. *Einstein: His Life and Universe*. New York: Simon and Schuster.

Jablonka, E., and M. Lamb. 2005. *Evolution in Four Dimensions: Genetic, Epigenetic, Behavioral, and Symbolic Variation in the History of Life*. Cambridge, MA: The MIT Press.

Jasanoff, S. 2005. 'Science and Environmental Citizenship' in P. Dauvergne, ed., *Handbook of Global Environmental Politics*. Cheltenham, UK: Edward Elgar, pp. 365–82.

Jaschik, S. 1990. 'Report Says NIH Ignores Own Rules on Including Women in Its Research', *Chronicle of Higher Education* June 27: A27.

Jarrell, A., and N. Ball. 1980. *Science, Technology and Canadian History*. Waterloo, ON: Wilfrid Laurier University Press.

Johnson, A. 1999. 'Still Platonic After all these Years: Artificial Life and the Form/Matter Dualism', *Australian Feminist Studies* 14 (29): 47–61.

———. 2009. 'Communicating the North: Scientific Practice and Canadian Postwar Identity', *The History of Science Society*, 24 (3) pp. 144–64.

Johnson, K. 2009. 'Climate Emails Stoke Debate: Scientists' Leaked Correspondence Illustrates Bitter Feud Over Global Warming', *The Wall Street Journal*, 23 November, p. A3.

Joly, P.B., and A. Kaufman. 2008. 'Lost in Translation? The Need for "Upstream"

Engagement" with Nanotechnology on Trial', *Science as Culture* 17 (3): 225–47.

Jones-Imhotep, E. 2004. 'Nature, Technology, and Nation', *Journal of Canadian Studies* 38 (3): 5–36.

———. 2005. 'Laboratory Cultures', *Scientia Canadensis: Canadian Journal of the History of Science, Technology and Medicine* 28: 6–26.

Joyce, K. 2005. 'Magnetic Resonance Imaging and the Production of Authoritative Knowledge', *Social Studies of Science* 35 (3): 437–62.

Kalof, L., ed. 2007. *The Animals Reader: The Essential Classical and Contemporary Writings*. New York: Berg.

Kant, I. 1781/2003. *The Critique of Pure Reason*. Mineola, NY: Dover Publications.

———. 1756/1994. *History and Physiography of the Most Neglected Cases of the Earthquake which Towards the End of the Year 1755 Shook a Great Part of the Earth* in S. Palmquist (ed) Four Neglected Essays by I. Kant. Hong Kong: Philopsychy Press, pp. 2–30.

Karter, A. 2003. 'Commentary: Race, Genetics, and Disease—in Search of a Middle Ground', *International Journal of Epidemiology* 32: 26–8.

Kearnes, M. 2003. 'Geographies That Matter: The Rhetorical Deployment of Latour, B.', *We Have Never Been Modern*, Catherine Porter, trans. Cambridge, MA: Harvard University Press.

Keller, E. Fox 1977. 'The Anomaly of a Woman in Physics', in S. Ruddick and P. Daniels, eds, *Working It Out*. New York: Pantheon Books, pp. 23–30.

———. 1983. *A Feeling for the Organism: The Life and Work of Barbara McClintock*. San Francisco: W.H. Freeman.

———. 1985. *Reflections on Gender and Science*. London: Yale University Press.

———. 2000. *The Century of the Gene*. Cambridge, MA: Harvard University Press.

———. 2002. *Making Sense of Life: Explaining Biological Development with Models, Metaphors and Machines*. Cambridge, MA: Harvard University Press.

Kidd, K.K. et al. 2006. 'Developing a SNP Panel for Forensic Identification of Individuals', *Forensic Science International* 164 (1): 20–32.

Kirby, V. 2006. *Judith Butler: Live Theory*. New York and London: Continuum.

Kitzinger, J. 2008. 'Questioning Hype, Rescuing Hope? The Hwang Stem Cell Scandal and the Reassertion of Hopeful Horizons', *Science as Culture* 17: 417–34.

Knorr Cetina, K. 1981. *The Manufacture of Knowledge: An Essay on the Constructivist and Contextual Nature of Science*. Oxford: Pergamon Press.

Kohler, R.E. 1994. *Lords of the Fly: Drosophilia Genetics and the Experimental Life*. Chicago: University of Chicago Press.

Kral J.G., et al. 2006. 'Large Maternal Weight Loss from Obesity Surgery Prevents Transmission of Obesity to Children Who were Followed for 2 to 18 Years', *Pediatrics* 118 (6): e1644–9.

Kuhn, T.S. [1962] 1970. *The Structure of Scientific Revolutions*. Chicago: University of Chicago Press.

Kumar, D. 2006. *Science and the Raj: A Study of British India*, 2nd edn. Oxford: Oxford University Press.

Lakatos, I., and A. Musgrave, eds. 1970. *Criticism and Growth of Knowledge*. Cambridge: Cambridge University Press.

Lancaster, J. 1989. 'Women in Biosocial Perspective', in S. Morgan, ed., *Gender and Anthropology*. Washington, DC: American Anthropological Association, pp. 95–115.

———. 1991. 'A Feminist and Evolutionary Biologist looks as Women', *Yearbook of Physical Anthropology* 34: 1–11.

Landecker, H. 2007. *Culturing Life: How Cells Became Technologies*. Cambridge, MA: Harvard University Press.

Lander, E.S. 1995. 'Mapping Heredity: Using Probabilistic Models and Algorithms to Map Genes and Genomes', in E.S. Lander and M.S. Waterman, eds, *Calculating the Secrets of Life: Applications of the Mathematical Sciences in Molecular Biology*. Washington DC: National Academy Press, pp. 25–55.

Lang, H.S. 1992. *Aristotle's Physics and its Medieval Varieties*. Albany: State University of New York Press.

Langton, M. 1999. 'The Fire That is the Centre of Each Family: Landscapes of the Ancients', in A. Hamblin, ed., *Visions of Future Landscapes*. Fenner Conference on the Environment 2.5. Canberra: Proceedings of the Australian Academy of Science.

Lassen, I. 2008. 'Commonplaces and Social Uncertainty: Negotiating Public Opinion', *Journal of Risk Research* 11 (8): 1025–45.

Latour, B. 1983. 'Give Me a Laboratory and I Will Raise the World', in K. Knorr Cetina and M. Mulkay, eds, *Science Observed: Perspectives on the Social Study of Science*. London: Sage, pp. 141–70.

———. 1987. *Science in Action: How to Follow Scientists and Engineers through Society*. Cambridge, MA: Harvard University Press.

————. 1988. *The Pasteurization of France*, A. Sheridan and J. Law, trans. Cambridge, MA: Harvard University Press.

————. 1993. *We Have Never Been Modern*, C. Porter, trans. New York: Harvester Wheatsheaf.

————. 1999. *Pandora's Hope: Essays on the Reality of Science Studies*. Cambridge, MA: Harvard University Press.

Latour, B., and S. Woolgar. 1986. *Laboratory Life: The Construction of Scientific Facts*. Sage: London.

Law, J. 1986. *Power, Action and Belief: A New Sociology of Knowledge?* London: Routledge.

————. 2008. 'On Sociology and STS', *The Sociological Review* 56 (4): 623–49.

Lawrence, A. 1991. 'Mother-daughter Bonds in Sheep', *Animal Behavior* 42: 683–5.

Lee, A. Forthcoming. 'Objectivity in Health Science—Case Study of the Cochrane Collaboration', PhD Dissertation, Queen's University.

Leedale, G. 1974. 'How Many are the Kingdoms of Organisms?', *Taxon* 23: 261–70.

Leplin, J. 1984. *Scientific Realism*. Berkeley, CA: University of California Press.

Leveque, J. 2008. 'Alien/Nature: Observations on the Relationship between Humanity, Nature, and Capital', MA thesis, Department of Political Studies, Queen's University.

————. Forthcoming. 'Our Brave New Future: Studying the Socio-Ecological Impacts of Genetic Technologies', PhD Dissertation, Queen's University.

Levere, J. 1974. *A Curious Field Book: Science and Society in Canadian History*. Toronto: Oxford University Press.

Levin, Y. 2008. *Imagining the Future: Science and American Democracy*. New York: Encounter Books.

Lewenstein, B. 1992. 'The Meaning of "Public Understanding of Science" in the United States after World War II', *Public Understanding of Science* 1: 45–68.

Lewenstein, B. 1995. 'From Fax to Facts: Communication in the Cold Fusion Saga', *Social Studies of Science*, 25: 403–36.

Lewontin, R.C. 2000. *The Triple Helix: Gene, Organism, and Environment*. Cambridge, MA: Harvard University Press.

Linder, D. 2004. *The Vatican's View of Evolution: The Story of Two Popes*. Available at www.law.umkc.edu/faculty/projects/ftrials/conlaw/vaticanview.html; accessed 29 March 2010.

Lingus, A. 1994. *Foreign Bodies*. New York: Routledge.

Lloyd, G. 1979. *Magic, Reason and Experience: Studies in the Origin and Development of Greek Science*. Cambridge: Cambridge University Press.

Lock, M. 1997. 'Decentering the Natural Body: Making Difference Matter', *Configurations* 5 (2): 267–92.

Longino, H.E. 1990. *Science as Social Knowledge: Values and Objectivity in Scientific Theory*. Princeton, NJ: Princeton University Press.

Luján, J., and O. Todt. 2007. 'Precaution in Public: The Social Perception of the Role of Science and Values in Policy Making', *Public Understanding of Science* 16: 97–109.

Lumey, L.H. et al. 2007. 'Cohort Profile: the Dutch Hunger Winter Families Study', *International Journal of Epidemiology* 36 (6): 1196–204.

Lupton, D., and W. Seymour. 2000. 'Technology, Selfhood and Physical Disability', *Social Science and Medicine* 50: 1851–62.

Lykke, N., and R. Braidotti. 1996. *Between Monsters, Goddesses, and Cyborgs: Feminist Confrontations with Science, Medicine, and Cyberspace*. London: Zed Books.

Lynch, M. 1985. *Art and Artifact in Laboratory Science: A Study of Shop Work and Shop Talk in a Research Laboratory*. London: Routledge.

Lynch, M., and S. Woolgar. 1990. *Representation in Scientific Practice*. Cambridge, MA: MIT Press.

MacDermot, H. 1967. *One Hundred Years of Medicine in Canada 1867–1967*. Toronto: McClelland and Stewart.

Mach, E. [1887] 2009. *Contributions to the Analysis of Sensations*. New York: General Books.

Mackenzie, A., and A. Murphie. 2008. 'The Two Cultures Become Multiple?: Sciences, Humanities and Everyday Experimentation', *Australian Feminist Studies* 23 (55): 87–100.

Maclaren, A. 1990. *Our Own Master Race: Eugenics in Canada, 1885–1990*. Toronto: McClelland and Stewart.

Malinowski, B. 1913. *The Family among Australian Aborigines*. London: University of London Press.

Margulis, L., and D. Sagan. *Slanted Truths: Essays on Gaia, Symbiosis, and Evolution*. New York: Copernicus.

Margulis, L., M. Dolan, and R. Guerrero. 1999. 'The Molecular Tangled Bank: Not Seeing the Phylogenies for the Trees', *Biology Bulletin* 196: 413–14.

Martin, E. 1990. 'The Egg and Sperm: How Science Has Constructed a Romance Based on Stereotypical Male-Female Roles', *Signs: Journal of Women in Culture and Society* 16 (3): 485–501.

Marx, K. 1964. *Economic and Philosophic Manuscripts of 1844*. New York: International Publishers.

Massey, D. 2005. 'Negotiating Nonhuman/ Human Place', *Antipode* 37 (2): 353–7.

Mayberry, M., B. Subramaniam, and L. Weasel, eds. 2001. *Feminist Science Studies: A New Generation*. New York: Routledge.

Mayer, R. 1997. *Inventing Canada: One Hundred Years of Innovation*. Vancouver: Raincoast Books.

Mayr, E. 2001. *What Evolution Is*. New York: Basic Books.

Medawar, P. 1963. 'Is the Scientific Paper a Fraud?', *The Listener* 70 (1798): 377–8.

Meillassoux, Q. 2008. *After Finitude: An Essay on the Necessity of Contingency*. New York: Continuum.

Meilwee, J., and J. Robinson. 1992. *Women in Engineering: Gender, Power, and Workplace Culture*. New York: SUNY Press.

Merchant, C. 1980. *The Death of Nature: Women, Ecology, and the Scientific Revolution*. San Francisco: Harper and Row.

Merton, R. 1973. *Sociology of Science: Theoretical and Empirical Investigations*. Chicago: University of Chicago Press.

Michael, M., and S. Carter. 2001. 'The Facts About Fictions and Vice Versa: Public Understanding of Human Genetics', *Science as Culture* 10 (1): 5–32.

Michaels, D., and C. Monforton. 2005. 'Manufacturing Uncertainty: Contested Science and the Protection of the Public's Health and Environment', *American Journal of Public Health* 95 (S1): S39–48.

Miller, J. 1991. *The Public Understanding of Science and Technology in the United States, 1990: A Report to the National Science Foundation*. DeKalb, IL: Public Opinion Laboratory, Northern Illinois University.

———. 2004. 'Public Understanding Of, and Attitudes Toward, Scientific Research: What We Know and What We Need to Know', *Public Understanding of Science* 13: 273–94.

Mills, C.W. 1959. *The Sociological Imagination*. Oxford: Oxford University Press.

Moe, K. 1984. *Should the Nazi Research Data be Cited?* Hastings Center Report, Institute of Society, Ethics, and the Life Sciences.

Moore, A., and J. Stilgoe. 2009. 'Experts and Anecdotes: The Role of "Anecdotal Evidence" in Public Scientific Controversies', *Science, Technology, and Human Values* 34 (5): 654–77.

Mooreg, M. 2009. 'Climate Change

Scientists Face Calls for Public Inquiry over Data Manipulation Claims', *The Daily Telegraph*, 24 November.

Mort, M., T. Finch, and C. May. 2009. 'Making and Unmaking Telepatients: Identity and Governance in New Health Technologies', *Science, Technology, and Human Values* 34 (1): 9–33.

Mortimer-Sandilands, C., and B. Erickson, eds. 2010. *Queer Ecologies: Sex, Nature, Politics and Desire*. Bloomington, IN: Indiana University Press.

Morton, D. 1990. *A Military History of Canada*. Edmonton: Hurtig.

Mosco, V. Forthcoming. 'Entanglements: Between Two Cultures and Beyond Science Wars', *Science as Culture*.

Moser, I. 2000. 'Against Normalisation: Subverting Norms of Ability and Disability', *Science as Culture* 9 (2): 201–40.

———. 2006. 'Sociotechnical Practices and Difference: On the Interferences between Disability, Gender, and Class', *Science, Technology, and Human Values* 31 (5): 537–64.

Murphie, A., and A. Mackenzie. 2008. 'The Two Cultures Become Multiple? Sciences, Humanities and Everyday Experimentation', *Australian Feminist Studies* 22 (5): 87–100.

Murphy, J. 1989. 'Is Pregnancy Necessary? Feminist Concerns about Ectogenesis', *Hypatia* 4 (3): 66–8.

Murphy, M. 2006. *Sick Building Syndrome and the Problem of Uncertainty*. Durham, NC: Duke University Press.

National Academy of Sciences. 1999. *Science and Creationism: A View from the National Academy of Sciences*, 2nd edn. Washington, DC: National Academy Press.

Needham, J., K. Robinson, and J.Y. Huang. 2004. *Science and Civilization in China. Volume 7, Part II: General Conclusions and Reflections*. Cambridge: Cambridge University Press.

Nelson, A. 2008. 'Genetic Genealogy Testing and the Pursuit of African Ancestry', *Social Studies of Science* 38 (5): 759–83.

Nelson, J. 2002. 'Microchimerism: Incidental Byproduct of Pregnancy or Active Participant in Human Health?', *Trends in Molecular Medicine* 8 (3): 109–13.

Nicosia, F., and J. Huener. 2002. *Medicine and Medical Ethics in Nazi Germany: Origins, Practice, Legacies*. New York: Berghahn Books.

Novas, C., and N. Rose. 2000. 'Genetic Risk and the Birth of the Somatic Individual', *Economy and Society* 29 (4): 485–513.

Nowotny, H. 1993. 'Socially Distributed

Knowledge: Five Spaces for Science to Meet the Public', *Public Understanding of Science* 2: 307–19.

Nowotny, H., P. Scott, and M. Gibbons. 2001. *Re-Thinking Science: Knowledge and the Public in an Age of Uncertainty*. Oxford: Polity Press.

Olausson, U. 2009. 'Global Warming-Global Responsibility? Media Frames of Collective Action and Scientific Certainty', *Public Understanding of Science* 18: 421–36.

Oliver, M. 1990. *The Politics of Disablement*. London: Macmillan.

O'Neill, T., and M.J. Hird. 2001. 'Double Damnation: Gay Disabled Men and the Negotiation of Masculinity', in K. Backett-Milburn and L. McKie, eds, *Constructing Gendered Bodies*. London: Palgrave, pp. 201–23.

Ottinger, G. 2010. 'Buckets of Resistance: Standards of the Effectiveness of Citizen Science', *Science, Technology, and Human Values* 35 (2): 244–70.

Oudshoorn, N. 1994. *Beyond the Natural Body. An Archeology of Sex Hormones*. London and New York: Routledge.

Overall, C. 1989. *The Future of Human Reproduction*. Toronto: The Women's Press.

Paddaiah, G. 2008. 'Human Genetics in the Service of Tribal Health', in V. Subramanyam, ed., *Indigenous Science and Technology for Sustainable Development*. Jaipur: Rawat Publications, pp. 216–24.

Parisi, L. 2004. *Abstract Sex: Philosophy, Biotechnology and the Mutations of Desire*. New York and London: Continuum.

Pasteur, L. 1854. *Lecture at University of Lille*. 7 December.

Pearce, F. 2010. 'Part Two: How the 'Climategate' Scandal is Bogus and Based on Climate Sceptics' Lies', *The Guardian*, 9 February.

Pearson, H. 2002. 'Dual Identities', *Nature* 417: 10–11.

Peirce, C.S. 1965. *Collected Papers*. Cambridge, MA: Harvard University Press.

Pellechia, M. 1997. 'Trends in Science Coverage: A Content Analysis of Three US Newspapers', *Public Understanding of Science* 6: 49–68.

Pellegrino, E., and N. Baumslag. 2005. *Murderous Medicine: Nazi Doctors, Human Experimentation, and Typhus*. New York: Praeger Publishers.

Pembrey M., et al. 2006. 'ALSPAC Study Team: Sex-specific, Male-line Transgenerational Responses in Humans', *European Journal of Human Genetics* 14 (2): 159–66.

Pestre, D. 2008. 'Challenges for the Democratic Management of Technoscience: Governance, Participation and the Political Today', *Science as Culture* 17 (2): 101–19.

Petersen, A., A. Anderson, and S. Allan. 2005. 'Science Fiction/Science Fact: Medical Genetics in News Stories', *New Genetics and Society* 24 (3): 337–53.

Pickering, A. 1999. *Constructing Quarks: A Sociological History of Particle Physics*. Chicago: University of Chicago Press.

———. 1995. *The Mangle of Practice: Time, Agency, and Science*. Chicago: University of Chicago Press.

Pickering, M. 1993. *Auguste Comte: An Intellectual Biography*. Cambridge: Cambridge University Press.

Pinker, S. 2002. *The Blank Slate: The Modern Denial of Human Nature*. New York: Penguin Books.

Plant, S. 1997. *Zeros and Ones*. London: Fourth Estate.

Polanyi, M. 1958. *Personal Knowledge: Towards a Post-critical Philosophy*. Chicago: University of Chicago Press.

Popli, R. 1999. 'Scientific Literacy for all Citizens: Different Concepts and Contents', *Public Understanding of Science* 8: 123–37.

Poudrier, J. 2007. 'The Geneticization of Aboriginal Diabetes and Obesity: Adding Another Scene to the Story of the Thrifty Gene', *The Canadian Review of Sociology and Anthropology* 44 (2): 237–61.

Powell, M., S. Dunwoody, R. Griffin, and K. Neuwirth. 2007. 'Exploring Lay Uncertainty about an Environmental Health Risk', *Public Understanding of Science* 16: 323–43.

Powell, R. 2008. 'Science, Sovereignty and Nation: Canada and the Legacy of the International Geophysical Year, 1957–1958', *Journal of Historical Geography* 34: 618–38.

Prakash, G. 1999. *Another Reason: Science and the Imagination of Modern India*. Princeton, NJ: Princeton University Press.

Proctor, R. 1990. *Racial Hygiene: Medicine under the Nazis*. Boston: Harvard University Press.

———. 2000. *The Nazi War on Cancer*. Princeton, NJ: Princeton University Press.

Protevi, J. 2008. *Three Lectures on Deleuze and Biology*; available at www.protevi.com/john/research.html

Pruetz, J., and P. Bertolani. 2007. 'Savanna Chimpanzees, *Pan troglodytes versus*, Hunt with Tools', *Current Biology* 17: 412–17.

Pyne, S. 2001. *Fire: A Brief History*. Seattle:

University of Washington Press.

Quine, W.V. 1960. *Word and Object*. Cambridge, MA: MIT Press.

Rackham, M. 2000. 'My Viral Lover', *Real-Time/OnScreen* 37 (June): 22.

Raffles, H. 2002. 'The Dreamlife of Ecology: South Pará, 1999', in *In Amazonia: A Natural History*. Princeton, NJ: Princeton University Press, pp. 150–79.

Raina, D., and S. Irfan Habib. 2004. *Domesticating Modern Science: A Social History of Science and Culture in Colonial India*. New Delhi: Tulika.

Reardon, J. 2005. *Race to the Finish: Identity and Governance in an Age of Genomics*. Princeton, NJ: Princeton University Press.

Remington, J. 1988. 'Beyond Big Science in America: The Binding of Inquiry', *Social Studies of Science* 18: 45–72.

Revkin, A. 2009. 'Hacked E-Mail Is New Fodder for Climate Dispute', *The New York Times*, 20 November.

Roberts, C. 2007. *Messengers of Sex: Hormones, Biomedicine and Feminism*. Cambridge: Cambridge University Press.

Roberts, D. 2008. 'Race, Gender, and Genetic Technologies: A New Reproductive Dystopia?', *Signs* 34 (4): 783–804.

Robinson, S. 2010. 'Epistemic Values and Epistemologies of the Eye: Lorraine Daston and Peter Galison's Objectivity', *Parallax* 16: 113–16.

Robinson, W. 1997. 'Some Nonhuman Animals Can Have Pains in a Morally Relevant Sense', *Biology and Philosophy* 12: 51–71.

Rose, N., and C. Novas. 2005. 'Biological Citizenship', in A. Ong and S. Collier, eds, *Global Assemblages: Technology, Politics, and Ethics as Anthropological Problems*. Malden, MA: Blackwell, pp. 439–63.

Rose, S., R. Lewontin, and L. Kamin. 1991. *Not in Our Genes: Biology, Ideology, and Human Nature*. London: Penguin Books.

Roseboom, T.J., et al. 2000. 'Coronary Heart Disease after Prenatal Exposure to the Dutch Famine, 1944–45', *Heart* 84 (6): 595–8.

Roseboom T.J., J.H. van der Meulen, C. Osmond, D.J. Barker, A.C. Ravelli, and O.P. Bleker. 2000. 'Plasma Lipid Profiles in Adults after Prenatal Exposure to the Dutch Famine', *American Journal of Clinical Nutrition* 72 (5): 1101–6.

Rossiter, M.W. 1984. 'Women Scientists in America', *Bulletin of the American Academy of Arts and Sciences* 36 (6): 10–16.

Roulstone, A. 1998. 'Researching a Disabling Society: The Case of Employment and New Technology', in T. Shakespeare, ed., *The Disability Reader: Social Science Perspectives*. London: Cassell, pp. 110–28.

Rouse, J. 1996. *Engaging Science: How to Understand its Practices Philosophically*. Ithaca, NY: Cornell University Press.

Rowe, G., T. Horlick-Jones, J. Walls, and N. Pidgeon. 2005. 'Difficulties in Evaluating Public Engagement Initiatives: Reflections on an Evaluation of the UK *GM Nation?* Public Debate about Transgenic Crops', *Public Understanding of Science* 14: 331–52.

Rowell, T. 1974. 'The Concept of Social Dominance', *Behavioral Biology* 11: 131–54.

———. 1979. 'How Would We Know if Social Organization Were Not Adaptive?', in I. Bernstein and E. Smith, eds, *Primate Ecology and Human Origins*. New York: Garland, pp. 1–22.

———. 1984. 'Introduction: Mothers, Infants and Adolescents', in M. Small, ed., *Female Primates*. New York: Alan Liss, pp. 13–16.

———. 1991. 'Till Death Do Us Part: Longlasting Bonds between Ewes and Their Daughters', *Animal Behavior* 42: 681–2.

———. 1993. 'Reification of Social Systems', *Evolutionary Anthropology* 2 (4): 135–7.

Salazar, M., and A. Holbrook. 2007. 'Canadian Science, Technology and Innovation Policy: The Product of Regional Networking?', *Regional Studies* 41 (8): 1129–41.

Saldanha, A. 2006. 'Reontologizing Race: The Machinic Geography of Phenotype', *Environment and Planning D: Society and Space* 24: 9–24.

Sankar, P. 2006. 'Hasty Generalisation and Exaggerated Certainties: Reporting Genetic Findings in Health Disparities Research', *New Genetics and Society* 25 (3): 249–64.

Satzewich, V., and N. Liodakis. 2010. *'Race' and Ethnicity in Canada*, 2nd edn. Toronto: Oxford University Press.

Sawicki, J. 1999. 'Disciplining Mothers: Feminism and the New Reproductive Technologies', in J. Price and M. Shildrick, eds, *Feminist Theory and the Body*. Edinburgh: Edinburgh University Press.

Schäfer, M. 2009. 'From Public Understanding to Public Engagement: An Empirical Assessment of Changes in Science Coverage', *Science Communication* 30 (4): 475–505.

Schickore, J. 2007. *The Microscope and the Eye: A History of Reflections 1740–1870*. Chicago: University of Chicago Press.

Schiebinger, L. 1989. *The Mind has No Sex?*

Women in the Origins of Modern Science. Cambridge, MA: Harvard University Press.

———. 1993. *Nature's Body: Gender in the Making of Modern Science.* Boston: Beacon Press.

Schneider, D. 1968. *American Kinship: A Cultural Account.* Englewood Cliffs, NJ: Prentice Hall.

———. 1980. *American Kinship: A Cultural Account,* 2nd edn. Chicago: University of Chicago Press.

Schrader, A. 2006. 'Phantomatic Species Ontologies: Questions of Survival in the Remaking of Kin and Kind', 4S Conference, November, Vancouver, BC, pp. 1–10.

Schwartz, R. 2001. 'Racial Profiling in Medical Research', *New England Journal of Medicine* 344 (18): 1392–93.

Sclove, R. 1996. 'Town Meetings on Technology', *Technology Review* 99 (5): 24–31.

———. 2001. 'STS on Other Planets', in S. Cutcliffe and C. Mitcham, eds, *Visions of STS: Counterpoints in Science, Technology, and Society Studies.* New York: State University of New York Press, pp. 111–21.

Scott, R. Forthcoming. 'Laboratory Lives of Afterbirths: Placentas as Working Objects of Study', PhD thesis, Queen's University.

Serres, M. 1995. *The Natural Contract,* E. MacArthur and W. Paulson, trans. Ann Arbor, MI: University of Michigan Press.

Shapin, S. 1988. 'Understanding the Merton Thesis', *Isis* 79 (4): 594–605.

Shapin, S., and S. Schaffer. 1985. *Leviathan and the Air Pump: Hobbes, Boyle and the Experimental Life.* Princeton, NJ: Princeton University Press.

———. 2010. *Never Pure: Historical Studies of Science as if It Was Produced by People with Bodies, Situated in Time, Space, Culture.* Hopkins Fulfillment Service. **Material excerpted from Hackett, Edward J., Olga Amsterdamska, Michael E. Lynch, and Judy Wajcman, eds., *The Handbook of Science and Technology Studies,* third edition, pp. 433–448, © 2007 Massachusetts Institute of Technology, by permission of the MIT Press.

Sismondo, S. 2004. *An Introduction to Science and Technology Studies.* London: Blackwell Publishers.

Skolimowski, H. 1992. *Living Philosophy: Eco-Philosophy as a Tree of Life.* New York: Arkana Books.

Small, M., ed. 1984. *Female Primate: Studies by Female Primatologists.* New York: Liss.

Smuts, B. 2001. 'Encounters with Animal Minds', *Journal of Consciousness Studies* 8 (5–7): 293–309.

———. 2007. *Sex and Friendship in Baboons.* Piscataway, NJ: Transaction Press.

Smylie, J., C. Martin, N. Kaplan-Myrth, L. Steele, C. Tait, and W. Hogg. 2003. 'Knowledge Translation and Indigenous Knowledge', *Circumpolar Health*: 139–143.

Sokal, A.D. 1996. 'Transgressing the Boundaries: Toward a Transgressive Hermeneutics of Quantum Gravity', *Social Text* 46/47: 217–52.

———. 2010. *Beyond the Hoax: Science, Philosophy and Culture.* Oxford: Oxford University Press.

Sokal, A.D. and Bricmont, J. 1999. *Fashionable Nonsense: Postmodern Intellectuals' Abuse of Science.* New York: Picador.

Soper, K. 1995. *What is Nature?* Oxford and Cambridge, MA: Blackwell.

Spallone, P., and D. Steinburg. 1987. *Made to Order: The Myth of Reproductive and Genetic Progress.* New York: Pergamon Press.

Spence, A., and E. Townsend. 2006. 'Examining Consumer Behavior toward Genetically Modified (GM) Food in Britain', *Risk Analysis* 26 (3): 657–70.

Spencer, H. [1851] 1888. *Social Statistics: Or, the Conditions Essential to Human Happiness Specified, and the First of them Developed.* New York: D. Appleton and Company.

Steffens, B. 2006. *Ibn al-Haytham: First Scientist.* New York: Morgan Reynolds Publishing.

Stengers, I. 1997. *Power and Invention: Situating Science.* Minneapolis, MN: University of Minnesota Press.

———. 2000. 'Another Look: Relearning to Laugh', *Hypatia* 15 (4): 41–54.

Stern, M. 2006. 'Dystopian Anxieties versus Utopian Ideals: Medicine from *Frankenstein* to *The Visible Human Project* and *Body Worlds*', *Science as Culture* 15 (1): 61–84.

Stocking, S.H., and L. Holstein. 2009. 'Manufacturing Doubt: Journalists' Roles and the Construction of Ignorance in a Scientific Controversy', *Public Understanding of Science* 18: 23–42.

Sturgis, P., H. Cooper, and C. Fife-Schaw. 2005. 'Attitudes to Biotechnology: Estimating the Opinions of a Better-Informed Public', *New Genetics and Society* 24 (1): 31–56.

Subbiah, S. 2004. 'Reaping What They Sow: The Basmati Rice Controversy and Strategies for Protecting Traditional Knowledge', *Boston College International and Comparative Law Review* 27 (2): 529–59.

Subramanian, B. 2009. 'Moored Metamorphoses: A Retrospective Essay on

Feminist Science Studies', *Signs: The Journal of Women in Culture and Society* 34 (4): 951–80.

Thomas, C. 2007. *Sociologies of Disability and Illness: Contested Ideas in Disability Studies and Medical Sociology.* Hounmills, Basingstoke: Palgrave Macmillan.

Thompson, C. 2001. 'Strategic Naturalizing: Kinship in an Infertility Clinic' in S. Franklin and S. McKinnon, eds, *Relative Values: Reconfiguring Kinship Studies.* Durham, NC: Duke University Press, pp. 175–202.

Thompson, W.I. 1991. 'Introduction', in W.I. Thompson, ed., *Gaia 2: Emergence: The New Science of Becoming.* Hudson, NY: Lindisfarne Press, pp. 11–29.

Thomson, M. 1978. *The Beginning of the Long Dash: A History of Timekeeping in Canada.* Toronto: University of Toronto Press.

Thomson, R. 1997. *Extraordinary Bodies: Figuring Physical Disability in American Culture and Literature.* New York: Columbia University Press.

Tontonoz, M. 2008. 'The Scopes Trial Revisited: Social Darwinism versus Social Gospel', *Science as Culture* 17 (2): 121–43.

Townsend, E., D. Clarke, and B. Travis. 2004. 'Effects of Context and Feelings on Perceptions of Genetically Modified Food', *Risk Analysis* 24 (5): 1369–84.

Trafzer, C., W. Gilbert, and A. Madrigal. 2008. 'Integrating Native Science into a Tribal Environmental Protection Agency (EPA)', *American Behavioral Scientist* 51 (12): 1844–66.

Traweek, S. 1988. *Beamtimes and Lifetimes: The World of High Energy Physicists.* Cambridge, MA: Harvard University Press.

Trenn, T. 1978. 'Thoruranium (U-236) as the Extinct Natural Parent of Thorium: The Premature Falsification of an Essentially Correct Theory', *Annals of Science* 35: 581–97.

———. 1980. 'The Phenomenon of Aggregate Recoil: The Premature Acceptance of an Essentially Incorrect Theory', *Annals of Science* 37: 81–100.

Tschanz, D.W. 2003a. 'Arab Roots of European Medicine', *Heart Views: The Official Journal of the Gulf Heart Association* 4 (2): 69–80.

———. 2003b. 'Ibn-Sina: "The Prince of Physicians"', *Journal of the International Society for the History of Islamic Medicine* 1: 47–9.

Tugwell, P., V. Robinson, J. Grimshaw, and N. Santesso. 2006. 'Systematic Reviews and Knowledge Translation', *Bulletin of the World Health Organization* 84 (8): 643–702.

Tutton, R. 2007. 'Constructing Participation in Genetic Databases', *Science, Technology and Human Values* 32 (2): 172–95.

Tyler, K. 2009. 'Whiteness Studies and Laypeople's Engagements with Race and Genetics', *New Genetics and Society* 28 (1): 37–50.

Vale, T., ed. 2002. *Fire, Native Peoples, and the Natural Landscape.* Washington: Island Press.

Väliverronen, E. 2006. 'Expert, Healer, Reassurer, Hero and Prophet: Framing Genetics and Medical Scientists in Television News', *New Genetics and Society* 25 (3): 233–47.

Van Loon, J. 2000. 'Parasite Politics: On the Significance of Symbiosis and Assemblage in Theorizing Community Formations', in C. Pierson and S. Tormey, eds, *Politics at the Edge.* New York: St. Martin's Press.

Van Wyck, P. 2002. 'The Highway of the Atom: Recollections along a Route', *Topia* 7: 99–113.

———. 2005. *Signs of Danger: Waste, Trauma, and Nuclear Threat.* Minneapolis, MN: University of Minnesota Press.

———. 2010. *The Highway of the Atom.* Montreal: McGill-Queen's University Press.

Vardalas, J., 2001. *The Computer Revolution in Canada: Building National Technological Competence.* Cambridge: The MIT Press.

Vardy, M. 2009. 'Metaphysical Closure and the Politics of Articulation of Climate Change', *Topia* 21: 29–40.

Vertesi, J. 2007. '"It's Too Red": The Construction of Visual Knowledge in the Mars Exploration Rover Mission'. Paper presented at the Society for the Social Study of Science Conference, Montreal.

von Weizsacker, C.F. 1980. *The Unity of Nature.* New York: Farrar, Straus and Giroux.

Waddington, C.H. 1953. 'Epigenetics and Evolution', in R. Brown and J. Danielli, eds, *Evolution.* Cambridge: Cambridge University Press, pp. 186–99.

Wallman, S. 1998. 'Ordinary Women and Shapes of Knowledge: Perspectives on the Context of STD and AIDS', *Public Understanding of Science* 7: 169–85.

Warner, K. 2008. 'Agroecology as Paricipatory Science: Emerging Alternatives to Technology Transfer Extension Practice', *Science, Technology, and Human Values* 33 (6): 754–77.

Waterlow, S. 1988. *Nature, Change, and Agency in Aristotle's Physics: A Philosophical Study.* Oxford: Clarendon Press.

Watson, J.D. 1968. *The Double Helix: A*

Personal Account of the Discovery of the Structure of DNA. New York: Atheneum.

Webster, B. and Leake, J. "Scientists in Stolen E-mail Scandal Hid Climate Data." *The Times.* Times Newspapers Ltd, 28 Jan 2010 : n.pag. Web. 20 April 2011.

Weindling, P. 2006. *Nazi Medicine and the Nuremberg Trials: From Medical War Crimes to Informed Consent.* New York: Palgrave Macmillan.

Weir, L. 1998. 'Pregnancy Ultrasound in Maternal Discourse', in M. Shildrick and J. Price, eds, *Vital Signs.* Edinburgh: Edinburgh University Press, pp. 78–101.

Wertheim, M. 1995. *Pythagoras' Trousers: God, Physics, and the Gender Wars.* New York: Random House.

Westfall, R.S. 1983. *Never at Rest: A Biography of Isaac Newton.* New York: Cambridge University Press.

Weston, K. 2001. 'Kinship, Controversy, and the Sharing of Substance: The Race/ Class Politics of Blood Transfusion', in S. Franklin and S. McKinnon, eds, *Relative Values: Reconfiguring Kinship Studies.* Durham, NC: Duke University Press, pp. 147–74.

Whatmore, S. 2009. 'Mapping Knowledge Controversies: Science, Democracy, and the Redistribution of Expertise', *Progress in Human Geography* 33 (5): 587–98.

Whewell, W. 1837. *History of the Inductive Sciences: From the Earliest to the Present Time.* London: John W. Parker, West Strand.

Whitehead, A.N. [1920] 2007. *The Concept of Nature.* New York: Cosimo Classics.

Whitmarsh, L. 2009. 'What's in a Name? Commonalities and Differences in Public Understanding of "Climate Change" and "Global Warming"', *Public Understanding of Science* 18: 401–20.

Whittaker, R. 1969. 'New Concepts of Kingdoms of Organisms', *Science* 163: 150–60.

Whittaker, R., and L. Margulis. 1976. 'Protist Classification and the Kingdoms of Organisms', *BioSystems* 10: 3–18.

Widdowson, F., and A. Howard. 2008. *Disrobing the Aboriginal Industry: The Deception behind Indigenous Cultural Preservation.* Montreal and Kingston: McGill-Queen's University Press.

Wilson, E. 2004. *Psychosomatic: Feminism and the Neurological Body.* Durham, NC: Duke University Press.

Wilson, E.O. 1975. *Sociobiology: The New Synthesis.* Boston: Harvard University Press.

Winance, M. 2006. 'Trying Out the Wheelchair: The Mutual Shaping of People and Devices through Adjustment', *Science, Technology, and Human Values* 31 (1): 52–72.

Winant, H. 2004. *The New Politics of Race: Globalism, Difference, Justice.* Minneapolis, MN: University of Minnesota Press.

Witz, A. 2000. 'Whose Body Matters? Feminist Sociology and the Corporeal Turn in Sociology and Feminism', *Body and Society* 6 (2): 1–24.

Woese, C., and G. Fox. 1977. 'Phylogenetic Structure of the Prokaryotic Domain: The Primary Kingdoms', *Proceedings of the National Academy of Sciences USA* 74 (11): 5088–90.

Woese, C., O. Kandler, and M. Wheelis. 1990. 'Towards a Natural System of Organisms: Proposal for the Domains Archaea, Bacteria, and Eucarya', *Proceedings of the National Academy of Sciences USA* 87: 4576–9.

Woolgar, S. 1992. 'Some Remarks about Positivism: A Reply to Collins and Yearley', in Andrew Pickering, ed., *Science as Practice and Culture.* Chicago: University of Chicago Press, pp. 327–42.

Wright, J.W. 1991. *A History of the Native Peoples of Canada: Volume I.* Ottawa: Canadian Museum of Civilization.

———. 2001. *A History of the Native Peoples of Canada: Volume II.* Ottawa: Canadian Museum of Civilization.

Wynne, B. 1992. 'Misunderstood Misunderstanding: Social Identities and Public Uptake of Science', *Public Understanding of Science* 1: 281–304.

———. 1996. 'May the Sheep Safely Graze? A Reflexive View of the Expert-Lay Knowledge Divide', in S. Lash, B. Szerszynshi, and B. Wynne, eds, *Risk, Environment and Modernity.* London: Sage, pp. 44–83.

Yardley, L. 1997. 'The Quest for Natural Communication: Technology, Language and Deafness', *Health* 1 (1): 37–56.

Yibarbuk, D. 1998. 'Notes on Traditional Use of Fire on Upper Cadell River', in M. Langton, ed., *Burning Questions: Emerging Environmental Issues for Indigenous Peoples in Northern Australia.* Darwin: Centre for Indigenous Natural and Cultural Resource Management, Northern Territory University.

Zack, N. 2002. *Philosophy of Science and Race.* New York: Routledge.

Zihlman, A. 1985. 'Gathering Stories for Hunting Human Nature', *Feminist Studies* 11: 364–77.

Index

11, 13

Native peoples of Canada: *see* first peoples of Canada; *see also* indigenous peoples; indigenous science

natural philosophers (precursors of scientists), 4, 6, 8, 10, 14, 16, 17, 24

Natural Sciences and Engineering Research Council of Canada (NSERC), 13

natural selection, Darwin's theory of, 16

nature: domination/control of, 9, 10, 14, 89, 93–4; and separation from culture, 44, 59, 95–6; and society, 42–3

Netherlands: university programs for public in, 124; village famine in, and origin of epigenetics, 64–5

Networks of Centres of Excellence of Canada (NCE), 13

Newton, Isaac: biography of, 73; and heliocentrism, 33, 34; and measure of gravitational force, 77; and order/reality of mathematics, 31

Newtonian physics (classical physics/mechanics), 31, 41, 52, 53–4, 65–6; closed-environment study of, 53–4, 66; as reductionist, 52, 54, 59, 66; and Wave-Particle Duality Paradox, 56, 56–7

NitroMed (pharmaceutical company), 104

Nobel Prize winners, 11; for DNA research, 61, 92; McClintock, 63, 69, 92, 93; Penzias and Wilson, 72

nonhuman entities, 1, 37, 43, 44–9; Actor Network Theory and, 44–5, 49, 65, 66, 69; feminist studies and, 110; mangling and, 46–8, 49, 65, 66; Marxist view of, 37; SSK and, 43, 45, 48–9, 69

non-western civilizations: kinship among, 128; science and technology of, 5, 6–8, 20, 105

'normal science,' 40–1, 48, 77, 80

norms and values of science, 115–16, 133

Northwest Passage, 133

nucleotide replacement, 63

Nuremberg Doctors' Trial, 114

obesity, 65, 101

Occam's Razor, 54

occupational health and safety, Nazi initiatives in, 115

OncoMouse™, 132

Optical Methods in Biology, 78–9

optics, 7; and diffraction, 78

organic diet, Nazi promotion of, 115

Ottinger, Gwen, 127

Oxford University, 36, 123

Pablum, 11

pacemaker, artificial, 12

paradigms, 39–41, 42, 48; and anomalies, 72; and health research, 101; and 'normal science,' 40–1, 48, 80; and pedagogy, 38–40, 48; and scientific controversies, 120; and scientific method, 73, 79–80

paradigm shifts, 40; and quantum theory, 41, 52, 65

Parlby, Irene, 97

Pasteur, Louis, 76; Latour's study of, 73–4

Pearl Harbor, Japanese attack on, 22

pedagogy and paradigms, 38–40, 48

peer review, 29, 82, 116, 121

Peirce, Charles Sanders, 60

Penzias, Arno, 72

Perimeter Institute (Waterloo, Ont.), 13

Persia and origins of modern science, 7

Persons Case, 97

Perutz, Max, 92

Pestre, Dominique, 127

Pfiesteria piscicida, 77

phenomena: in air-pump experiment, 35, 81; as created/studied in laboratory, 70, 71, 76, 77, 83; as DNA markers of diseases/conditions, 63; in 'mangle' concept, 46–7; and models, 76; natural, 17, 19; reality as, 58–9; social construction of, 60; study of, 31; technology as, 107

phenomenology, 31

phenotypes, 61, 64; of race, 97, 98, 99

Philip Morris company, 125

'photograph 51,' 92

physics, 4, 11, 12, 22, 24, 34, 42; atom and, 61; closed experiments in, 53–4; Newtonian, 31, 41, 52, 53–4, 56–7, 59, 65–6; vs. philosophy, 53; *see also* quantum theory

Pickering, Andrew, 46–7, 49, 58, 79, 80, 83

placenta research, 81

Planck, Max, 22

Plato: Academy of, 7; *Dialogues* of, 30–1; and physics vs. philosophy, 53; and theory before method, 71

plenism, 34

Pliny, 6

Pluto, classification of, 120

Podolsky, Boris, 57, 58

polio, 11, 12

political economies of science and technology, 37–8, 53, 88, 89–91

Pons, Stanley, 120

Pope, Alexander, 31

positivism, 53, 84–5; and Merton Thesis, 115–16

post-colonial science studies: *see* race; race and health issues

Poudrier, Jennifer, 103

power, and science, 21–3, 53, 88–110; and ability/disability, 107–9; in Enlightenment era, 9, 10, 14, 89; and feminism, 91–6; and funding, 12–13, 21–3, 89–91; and political economy, 89–91; and race, 96–107

pre-Socratics, 5

administration, 12
transposons, genetic, 63, 92, 93
Trenn, Thaddeus, 34
TRF(H) peptide, 76
Trobriand Island, peoples of, 128
Truman, Harry S., 22
trust, in science, 116, 117–18, 120–3, 134; among scientists, 122
truth and falsehood: of beliefs, 42, 48; and scientific realism, 30; of Wassermann test, 39
Tschanz, David, 7
Tuskegee Institute syphilis study, 102
typhus, 38

UCSC Genome Bioinformatics Group, 62
UK Biobank, 127
Uncertainty Principle, 54, 57, 58
understanding of science, by public, 118–20, 122–4; and deficit model, 118–19; as facilitated by universities, 124; factors determining, 119–20, 134; as gleaned from popular sources, 117, 119; and knowledge translation, 119–20, 122–3, 134
'undone science,' 124–5
universities: ancient, 8; in Canada, 10–13; disciplinary separations at, 17; and scientific research/programs, 1–2, 11–13, 21–2, 24, 89, 123
University of Alberta, 2
University of British Columbia, 2
University of East Anglia Climate Research Unit, 121
University of Montreal, 13
University of Saskatchewan, 13, 103
University of Toronto, 1; and insulin, 11; and polio vaccine, 12; stem cell research at, 13
University of Western Ontario, 13
uranium, 22, 34; mining of, in Canada, 23

vaccines/vaccination, 12; Latour's study of, 73–4
validity: controversies of, 120, 126; of models, 77; of Nazi experiments, 114; of scientific articles, 82; of scientific facts, 36, 70, 71, 115, 120, 126; see also reliability
values and norms, of science, 115–16, 133
van Wyck, Peter: Highway of the Atom, 23
Vardalas, John: The Computer Revolution in Canada: Building National Technological Competence, 2
Vertesi, Janet, 80

Victorian Order of Nurses, 97
Visible Human Project, 122
vitamin D, 11

war, and science, 11–12, 21–3
Wassermann, August von, and test for syphilis, 39
Watson, James, 60, 61, 92; autobiography of, 73
Wave-Particle Duality Paradox, 41, 54–7; Bohr's *gedanken* on, 56–7, *56–7*; and Copenhagen Interpretation, 57–8; and Maxwell's Demon, 55; and Newtonian physics, 56, 56–7; and quantum theory, 41, 54, 56–7, *57*; and Schrödinger's Cat, 54, 55
weapons of mass destruction, 128; see also atomic weaponry, development of
Weston, Kath, 131–2
Whatmore, Sarah, 123
What the Bleep? (film), 59
Whewell, William: *History of the Inductive Sciences*, 10
Whitmarsh, Lorraine, 119–20
Wilkins, Maurice, 60, 61, 73, 92
Wilmut, Ian, 81
Wilson, Elizabeth, 95–6
Wilson, R.W., 72
women, 91–6, 128; classification of, 120, 121; and concept of sex differences, 91, 94–5; and environment, 93–4; and health research, 101; and knowledge, 93; as 'persons,' 97; as scientists, 3, 59, 60, 63, 69, 73, 91–3; and separation of nature and culture, 95–6; and technology, 21, 91, 94
women, in science, 3, 59, 91–3; biographies/studies of, 63, 69, 92–3; and Franklin's experience, 60, 73, 92; and McClintock's experience, 63, 69, 91, 92, 93
Woodward, Henry, 11
World Wide Web, 108, 127
Wynne, Brian, 123

xenotransplantation, 132–3
X-rays, 35–6

York University, 1
Young, Thomas: two-slit wave-duality experiment of, 56–7, *56–7*

ZENN electric car, 13
zoology, 11, 20